Have you ever wondered. . .

* How the Pyramids and Stonel

* If the Russians can trigger earthquakes in the U.S.?

* Whether the alchemists really transmuted iron into gold?

* If time travel is really possible?

If scientists really wanted to, could they . . .

* Demonstrate the Theory of Relativity with a pinch of salt on the kitchen table?

* Draw free energy from the earth's atmosphere?

* Construct a flying craft that behaves just like a UFO?

ISBN 9781301447220

Your visit today is profoundly
appreciated. Betty Joe.
Thank you.
Tom Pawels
18/9/19

How to Build a Flying Saucer

And Other Proposals in Speculative Engineering

T.B. Pawlicki

CORGI BOOKS

HOW TO BUILD A FLYING SAUCER

A CORGI BOOK 0 552 99014 0

First publication in Great Britain

PRINTING HISTORY
Corgi edition published 1983

Copyright © 1981 by T.B. Pawlicki

This book is set in 10/11 California

Corgi Books are published by
Transworld Publishers Ltd.,
Century House, 61–63 Uxbridge Road,
Ealing, London W5 5SA

Made and printed in Great Britain by the
Guernsey Press Co. Ltd., Guernsey, Channel Islands.

Contents

Introduction

Before the modern era, there was so little codified knowledge in general circulation that a little extra private knowledge was a powerful thing, especially in the engineering trades. So it was common for tradesmen to band together in closed societies to keep their special power to themselves. In our time, however, no one can possibly know more than a fraction of all the codified knowledge, so power comes from distributing knowledge rather than hoarding it.

Science presents itself as a disinterested enterprise devoted to discovering how the universe unfolds. When scientists discovered that facts were marketable, however, the pursuit of truth and the power devolving from knowledge became organized into an industry. To earn a living, therefore, a scientist must accept a certain way of perceiving how the universe unfolds as reality, and be blind to alternative ways of unfolding. This is why radical and revolutionary discoveries are made so frequently by people working and playing outside the profession of their interest (would Albert Einstein have conceived the theory of relativity if he had not suffered an academic failure permitting him to make his calculations in his spare time while working in an unhurried office?) or by lifetime amateurs like Gregor Mendel (the first known discoverer of genetics), Stanford Ovshinsky (the high-school dropout who discovered amorphous transistors), and Immanuel Velikovsky (who defied union rules by picking up the materials of other trades).

From time to time a layman will make a naive observation of commonplace evidence disregarded by the professionals as too trivial to be worthy of their interest, not immediately profitable enough to justify investment, or politically dangerous to mention. Nevertheless, the simple facts must prevail over current beliefs. This book is the result of this kind of observation.

Each of these stories is the consequence of observations professional people are discouraged from making. All the ideas, except the concepts of relativity, were worked out during shop-talk with skilled tradesmen in the course of everyday employment. The carpenter holds in his square the mathematics that was the glory of Greece. The boat designer is applying the principles of the most advanced calculus when he plots the curves of a hull. The electrician has fitted his leisure van with the electronic devices Q invents for James Bond. A tradesman possesses knowledge equivalent to professional scientists, and usually possesses somewhat greater skills because he not only has to be able to understand what the executives want, he also has to make these skills work. Most mechanics feel themselves to be at a disadvantage among professional people because they don't know that they possess equivalent knowledge. Because mechanics often do not learn to express themselves verbally in the manner recommended in the *Chicago University Manual of Style*, professional people do not recognize mechanics' learning, either.

Before a professional forum, amateurs—mechanics—look like idiots. Nevertheless, from time to time an amateur makes a naive and commonplace observation of such compelling importance that nothing else matters: Scientific belief must give way before a simple truth, however crudely presented.

This series of essays is the work of a lifetime amateur. These essays are presented as examples of this kind of momentous truth. Each is based on facts you can verify for yourself by making the same observations, and then proceeds to speculation. Each of these stories describes realities contradicting common sense and contradicting the local authorities on matters of science.

When published articles describing these observations brought response from university people, I was surprised to learn that the essential propositions in this book are in agreement with the most advanced thinkers of our time. These radicals are so far out of the common ball park that their very existence is unknown outside circles exchanging private papers. They are of the kind including Erwin Schrödinger, Louis de Broglie, and Nikola Tesla, names that are recognized today although the essence of their thinking remains mysterious. This is not to say that they endorse everything written in these pages; they don't even agree completely with each other. I believe, however, that you will find the errors insignificant in detail compared to the general conception.

Most amazing of all is being informed that the visionary theorists scattered in isolated outposts beyond the pale of authorized science are not so isolated nor so visionary nor so theoretical as believed. There are also outposts manned by engineers working with less fantasy and more hardware to turn the facts disregarded by the universities into operational machinery.

Refusal to clothe the truth with fiction made it impossible to publish this material for ten years. When 'The Crystal Planet' was first submitted, pyramid power made no sense to anyone except a few alchemists. Within the next five years, pyramids became the hottest property since the hula hoop. Less than five years ago, one editor, insisting that the public be given facts instead of fiction, was forced to leave his job. When the article, 'How to Build a Flying Saucer' finally appeared in print, after two years of editorial hassles, two editions included waivers of responsibility in case the facts proved true! But the following year, that very article was selected at the MUFON (Mutual UFO Network) international conference in Mexico City for translation into Spanish and publication throughout Latin America. By the end of the seventies, these essays had found their way all around the world, passed from person to person through the mails, and had been reproduced by basement presses until they achieved the kind of circulation enjoyed by liberal Soviet poets. In time, a reply from a Soviet engineer researching radical means of locomotion found its way back through the postal grapevine of UFO freaks! It is a common joke that American antigravity research consists mainly of a budget assigned to the CIA to spy on what the Russians are doing; it seems that the Russians may be reading American science fiction to find out what can be done.

Most of these essays appeared initially as a series of articles in *Pursuit*, the quarterly subscription journal published by the Society for the Investigation of the Unexplained, with its present public service office at RFD 5, Gales Ferry, Connecticut 06335. Each article had to be cut to the editorial slot and still be self-contained. The geometric concepts developed with increasing comprehension and precision over the years.

To amend repetitions and contradictions and bring the material up to date, the articles have been rewritten for easier and more informative reading.

A hundred years ago, a speed of twenty-five miles per hour

was legally prohibited in the belief that such terrific velocity would endanger life. Fifty years ago, engineers were certain man could not possibly reach the Moon because no rocket fuel exploded with sufficient velocity to escape the Earth's gravity. The limitation of combustion velocity was overcome by the concept of multiple-stage launchers. Neil Armstrong walked on The Moon without man's making a single new discovery. All that was necessary was to think differently about what was already known. The discoveries described in these essays are already in operation on the weapons proving grounds. One of these days an entire population is going to be annihilated like the citizens of Hiroshima, who did not realize there could be anything to fear from a lone aircraft on a beautiful August morning. Massive destruction seems to be the only way people will accept a revolutionary concept whose time is now.

So if anyone out there really wants to believe in flying saucers, he will ask, 'How can we do this saucer thing?' You will see how easy it is when you want to solve the mystery; how easy it is once you accept an attitude of solving one problem at a time as you come to it! If you learn nothing else from this exercise, you will at least have had a lesson in how easy it is to construct, from the facts discarded by orthodoxy, a scientific theory no less tenable than the authorized version.

1 Megalithic Engineering

How to Build Stonehenge and the Pyramids with Bronze Age Technology

The most generally believed myth prevailing today is that the Earth is a colonial outpost of an extraterrestrial civilization. In recent prehistoric times, this civilization exercised its super-sophisticated technology by quarrying inconveniently massive stones, transporting them for inconvenient distances over some of the world's most inconvenient terrain, and then piling them into temples on the scale of mountains—their way of getting their rocks off, so to speak. Whether or not you accept the scenario sold by Erich Von Daniken, scientists and engineers agree that the Great Pyramids of Egypt and other prehistoric megalithic monuments around the world are constructed on such a colossal magnitude that there's no way ordinary human beings could have done it. Even today, the monuments at Abu Simbel had to be cut into pieces to move them a few hundred yards from the Nile waters rising behind the Aswan Dam because our heaviest machinery was not otherwise able to manage the task. A number of authors, professional archaeologists, and more or less talented amateurs have published calculations proving that the magnitude of the pyramids exceeds all the human resources known to be available for construction at the time.

When I address myself to the construction methods of the pyramid builders, I am forced to turn in my amateur card, since I worked for several years in heavy earth construction. My last employer handled the largest contracts undertaken in this part of the country at the time, and I was being trained to supervise construction of an industrial town site. Whenever I read a story about how the prehistoric megalithic monuments may have been erected, even though I cannot make an on-site job survey, I check all the measurements to see that they agree. Every carpenter who is given plans for three twelve-foot rooms on a thirty-six-foot foundation knows it can't be done.

11

Architects frequently forget to allow for the thickness of the walls because they never have to do the job and because the contractor pays for errors.

If you take the trouble to calculate the volume of a megalith from the dimension given by various writers, and multiply it by the average weight of stone (with a specific density three times water), you will find the most impressive magnitude of the prehistoric building industry is error in simple arithmetic. Even if we assume that the length and width of the megaliths are not multiplied by literary hype, they are reduced to human proportions by correct calculation. Although Stonehenge could have been built the way the British scientists staged it for television, it wouldn't have been. The operations are regarded as too dangerous by today's unionized workmen. The prehistoric laborers' workmens' compensation would never have approved the contract. Estimating the job as if my employer had been asked to bid on Stonehenge, I found that standard techniques known throughout the trade proved simple and economical. It is inconceivable for the prehistoric engineers to have done it in any other way. Megalithic construction was the *usual* method of building public works the world over. In the Stone age it had to be the most economical technique available, just as steel frame and reinforced concrete is today; and the methods had to be simple enough for gormless apprentices to follow with no more than the usual number of goofs. If scientists had asked tradesmen how megalithic monuments could be built, there never would have been any mystery.

The standard way to erect pylons would have made no scene for Charlton Heston in *The Ten Commandments*. The pylon is drawn to the site on a stoneboat; pallets have been traditional in the moving trades since prehistoric times. A thousand slaves driven by whips are not necessary, because a thirty-foot pylon with an average cross section of five square feet weighs less than fifteen tons. Three hundred slaves, or a dozen oxen, could handle it easily. Even if the ancients had unlimited slaves, they used beasts of burden because animals are more efficient; the primitive Eskimo knows enough to hitch a team of dogs to a sled.

In the socket prepared to receive the pylon, the bottom is laid with a slab of stone for a footing to prevent the pylon's shifting and toppling over in the course of time. The edges of the socket, where the base of the pylon overhangs, is buttressed with logs to resist the pressure of the stone when raised against it.

On the other side of the socket from the pylon, a pole A-frame

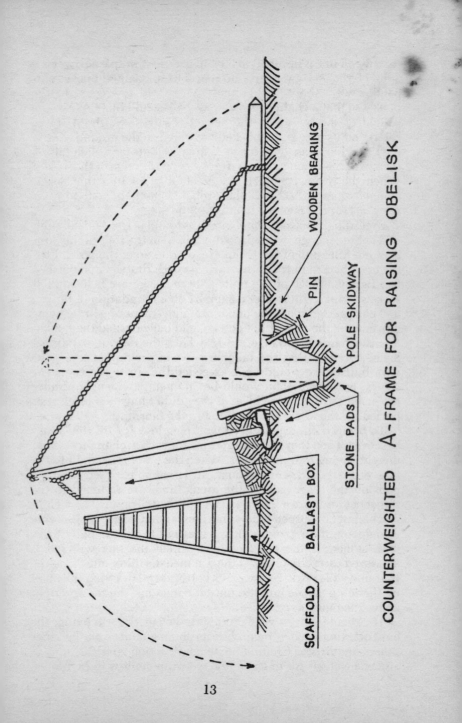

COUNTERWEIGHTED A-FRAME FOR RAISING OBELISK

WOODEN BEARING

PIN

POLE SKIDWAY

STONE PADS

BALLAST BOX

SCAFFOLD

is raised. A line is fastened to the top of the pylon, passed over the apex of the A-frame, and attached to an empty bucket suspended well above the ground. A scaffold is erected to the bucket so that a chain gang can pass ballast and fill it. When the weight of the ballast in the bucket is sufficient, the A-frame rotates on its legs as the bucket is lowered to the ground, drawing the pylon's base over the socket to the abutment, then raising the top of the pylon as the shaft rotates on the edge of the socket. When the pylon is practically erect, it plops into the socket. Voilà! No sweat. This is the probable method used to erect the posts for the massive triliths at Stonehenge.

Exceedingly massive lintels, such as found in the Andean ruins and at Stonehenge, were probably raised to the tops of the posts by a modification of the technique used to erect the posts. First the crew had to buttress the uprights with timbers and earthen fill. Barely a tenth of the posts at Stonehenge are held in their sockets; with so little lateral support on a foundation of broken and compressed rock, the horizontal movement of a lintel being moved over the tops of the posts would have pushed them over; this is a hardhat job. Gerald Hawkins, the decoder of Stonehenge, protests that the labor needed to provide all the posts with buttressing would have exceeded the effort of raising the lintels, but my platoon would be on unemployment insurance the day after they were hired if they didn't have a pair of posts buried for the arrival of the lintel in the morning.

To begin with, experienced workmen look for fill that's easy to shovel. Then they use the same fill for each arch in succession, merely having to fling it over. When the lintel is brought to the base of the buttress on its boat, an H-frame is raised on stone pads on the other side of the arch. Lines are fastened to the stoneboat and thrown over the legs of the H-frame. The weight of a ballast box suspended by the lines is what keeps the H-frame upright. A chain gang mounts a scaffold to fill the ballast box with stones. If the distance over which the box will fall is measured carefully to equal the distance the lintel must rise, the H-frame will slowly rotate on its bottom legs, drawing the lintel up the ramp on the buttress until it comes to a gentle stop right above the pins receiving it.

To skilled laborers who do this kind of work for a living, the only question is why the architects chose a quarry on the other side of the Bristol Channel. The construction specifications of Stonehenge call for about 100 five-ton monoliths to be hauled

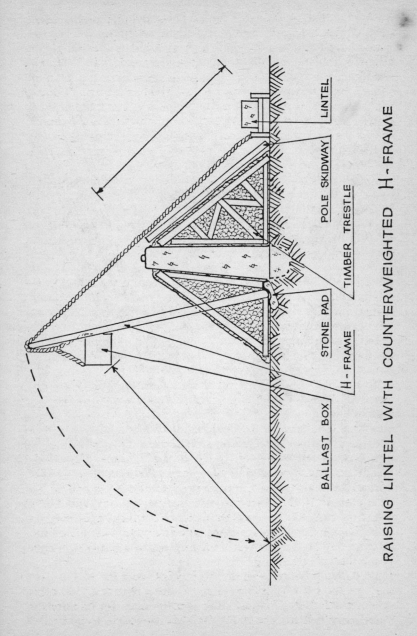

RAISING LINTEL WITH COUNTERWEIGHTED H-FRAME

LINTEL

POLE SKIDWAY

TIMBER TRESTLE

STONE PAD

H-FRAME

BALLAST BOX

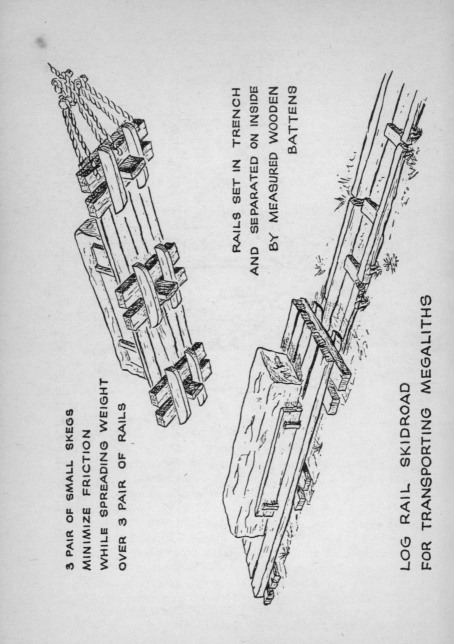

3 PAIR OF SMALL SKEGS
MINIMIZE FRICTION
WHILE SPREADING WEIGHT
OVER 3 PAIR OF RAILS

RAILS SET IN TRENCH
AND SEPARATED ON INSIDE
BY MEASURED WOODEN
BATTENS

LOG RAIL SKIDROAD
FOR TRANSPORTING MEGALITHS

twenty-five miles overland with a few hundred feet of rise. A five-ton load is no big drag for a couple yoke of oxen. The present structure at Stonehenge is the third of its kind built at the same site. The earlier stones were smaller and probably more numerous. By the time a few dozen sledges had been dragged over the trail from the quarry, the grade would have been cleared, soft spots filled with rocks, and slopes cut down to a comfortable angle. You need only see a path worn between farmhouse and outhouse to appreciate how quickly a repeated trail becomes a thoroughfare. To turn the grade into a proper railway, all that is needed is wooden poles laid along the right-of-way for the sledges to slide on. With some lubrication, properly cut skids will slide the stoneboats with less friction than the old Red River carts.

The smoothing of wooden rails for a skidway may seem more laborious than the dressing of the megaliths, but wood is ever so much easier to work. After one monument was completed, the roughly squared durable oak rails would be taken up and laid down for another construction elsewhere; the investment in transport machinery was thus amortized over many jobs. When the rails became crushed to the point where they could no longer bear the weight of sledges, the best sections—like worn-out railway ties—would be gleaned by poor folk for domestic structures, while the broken sections would be used for fuel.

Another shipment of about a hundred more blue stones had to be brought in from a quarry a few miles closer. To move these final megaliths, weighing from thirty to fifty tons, even twenty-mule teams would have sweated. To make the beasts sweat more heavily, between the quarry and the job site there is a low ridge to surmount. Archaeologists have sought the *easiest* grade over the ridge as the way the ancient Britons brought the megaliths to Stonehenge, which is why archaeologists cannot figure out how the monument was built. When there is a lot of altitude to gain in a short distance, however, experienced movers do not seek the easiest grade; they survey the range until they find the *steepest* cliff. Then they build a funicular railway, operated by counterweights, to raise the load as quickly as possible. (Travelers say that funicular transports on cliffsides are still used in England today.) Between the quarry and the job site there is a net loss of altitude, and once the cargo is raised to the summit of the dividing ridge, even fifty tons is no big sweat for teams of oxen to drag downhill.

Constructing the Great Pyramids was not a problem of engineering technique, but of logistics. The scale of the monuments is so awesome that it is pure fantasy to estimate on the contract without knowing the resources available.

Logistics aside, the pyramidal design expedites its own construction. To begin with, the pyramid is its own ramp. To raise the building blocks to the working level, all that needs to be done is to lay a pole skidway against the sides. A line is fastened to the stoneboat at the base of the pyramid and passed over a pulley wheel at the top level. The line is then strung right across the top level, passed over another pulley wheel on the opposite side, and fastened to another stoneboat carrying an empty ballast box. This stoneboat is held by the line on another pole skidway laid against that side of the pyramid.

Second, the stepped slope of the pyramid also serves as its own scaffolding. A chain gang standing on the steps passes small stones from the bottom of the pyramid to the ballast box on the top of the slope. When the box is filled, it slides down the skidway, dragging the building block up the other side. When the traverse is completed, both stoneboats are emptied, each load is replaced with its opposite number, and the transport is repeated in the opposite direction.

The Great Pyramid had sides four hundred feet long, so a dozen counterweighted shuttles would be in operation continuously, bringing blocks to the masons faster than they could be laid. As the pyramid rises, the area of working level diminishes; but at the same time, the number of blocks needed to lay each course is reduced by a factor of the square of the shortening of the sides, so the construction rises faster as it goes.

Dragging the pyramid blocks to working level up earthen ramps is one of several well publicized techniques worked out by scientists, including some professional engineers. Earthen ramps are plausible because they are feasible. But this pedestrian kind of engineering is never used by workmen in a competitive industry. An earthen ramp, besides requiring the moving of several times the amount of material involved in the pyramids themselves, won't stay put under long and heavy use. Earth flows under pressure, and the weight of the ramps themselves would be sufficient to generate flow. The earthen slopes of railway grades are built on a trestle framework filled with rock to keep them stabilized. It is impossible to walk on fresh earthen fill, nor will fresh fill support several tons of rock

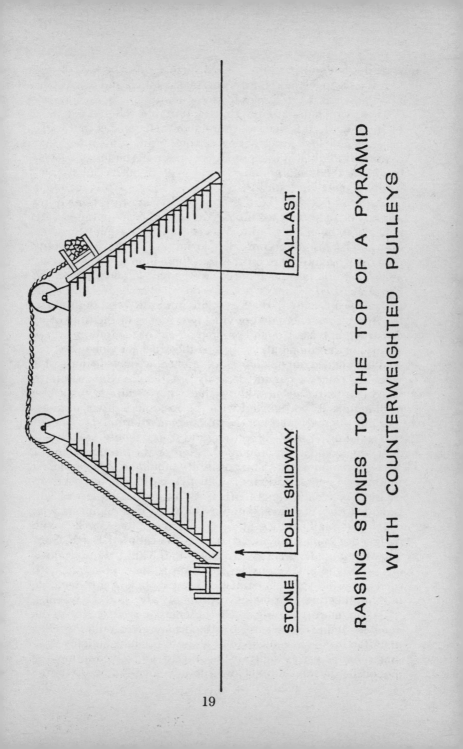

STONE | POLE SKIDWAY

BALLAST

RAISING STONES TO THE TOP OF A PYRAMID
WITH COUNTERWEIGHTED PULLEYS

and the slaves to haul it. A hard surface must be laid on the ramps to keep feet and load from sinking into it and tearing the ramp down about as fast as it can be rebuilt.

These problems are not insurmountable; all it takes is a lot more work. It is the cost-time-labor inefficiencies that make naive technology impractical. You must remember that pyramid building in ancient Egypt was a big industry, developed over thousands of years, and Egyptian businessmen practically invented accounting.

Critics object that there is no record of the Egyptians using wheels for industrial technology at the time the pyramids are believed to have been built. This is a specious argument when examined in historical context. The only reason we do not know how the pyramids were built is that there is no record to tell us. Without those missing records, we cannot know whether the Egyptians used wheels or not. We do not even know for certain that the people we call the Egyptians built the Great Pyramid in the first place. The only convincing evidence of the time of its construction is a cartouche painted on one of the interior stones signifying that the quarry supplied the order for Cheops.

When there is but one piece of evidence, we have to make of it what we can; we cannot deny its significance. But people in the construction business know that when a shipment of three million units is spot-labeled with the customer's order number, marking just one unit is not the practice; in a normal shipment of that amount, at least three thousand stones should be labeled. And why would a shipping label be left on the *interior* facing of a religious monument? The possibility should not be overlooked that the single scription of Cheops's cartouche may be an example of prehistoric graffiti. Until authenticated records are provided, it is not unreasonable to assume that engineers who developed mathematics at least as far as the Greeks, who designed a monument incorporating the ratios of Earth, Sun, and Moon, who perfected astronomy to the limit of the unaided eye, and who put a polished finish on a stone mountain with such precision that irregularities are practically undetectable without measurement by laser, probably also used the wheel.

Rather more puzzling is the absence of any record of the ancillary industries needed to support pyramid building. The manufacture of tool bits to cut stone must have employed as much manpower as the transport industry involved in moving monoliths by the ton. Construction men know that the replacement of lost tools is a serious expense for the company and crew. Yet not only is there no record of the means the prehistoric stonecutters used, but no tools have been found in the rubble left behind after construction. Neatness must have been rewarded beyond all other virtues in the ancient Egyptian construction industry; you need only consider the polished casing of the Pyramids to believe it.

Reporters compare the precise setting of the prehistoric megaliths with the casual tolerances of modern stonework and marvel. The only reason modern builders work to a tolerance of a quarter-inch is that there are no practical reasons to be more precise—and powerful economic reasons for not trying to be. (But if you ever visit a dry dock, you will see tonnages thousands of times more massive than the most stupendous works of the ancients set into position with a precision comparable to the Great Pyramid's; if a ship does not balance on its keel exactly where the blocks are set for it, it may topple over and put the dock out of business for several months until it is cut to pieces for removal like the monuments at Abu Simbel. Out of necessity, shipwrights developed techniques for placing hulls by hand because no machinery is big enough.) Given economic reason to be exacting, modern masons can equal the ancients. (The grinding of giant astronomical mirror lenses is a hand job). In a tradesman's opinion, the only reason the prehistoric megaliths were dressed to tolerances within a millimeter (at least once in a while) is that the labor was economical to perform within the context of the masons' operations.

A megalith brought into position on the course of stone being laid is supported on two leather bags filled with sand, one bag at each end. A platform of viscous clay or mortar is built up between the sandbags. When the bags are opened, the sand pours out, allowing the stone to settle onto the clay platform. As soon as the bags are free of the weight of the stone, they are snatched away by apprentices (the slow boys can be recognized kneading mortar with the stumps of their arms). If the top surface of the clay is slanted, the megalith will slide by its own weight, tight against its neighbor. The only technical problem is mixing clay to support the stone long enough for maneuvering, but still fluid enough to be squeezed out completely by the time the next stone is brought in for laying.

For a megalith weighing dozens of tons to slide easily over the clay, the broad bearing surfaces must be smoothly dressed. As

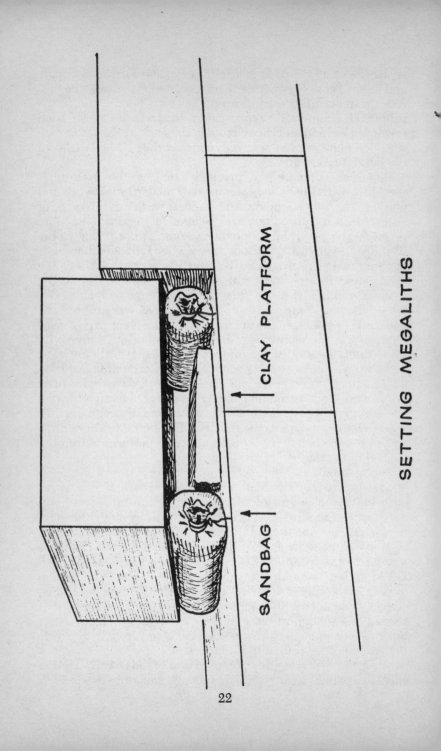

SANDBAG

CLAY PLATFORM

SETTING MEGALITHS

the smoothing of the broad surfaces cannot be left undone, it is no big deal to dress the abutting ends to an equal finish; the precision of the ends is not critical, so the job can be given to apprentices for their practice. A technique like this would be considered only when the cranes raising the stones to the working course were not adequate for precision setting.

It is likely that the millions of blocks used in the construction of pyramids were transported from the quarry to the job site over a rail skidway. Rails are far more efficient than log rollers. But once you see construction crews at work, other possibilities come to mind. We know that megaliths were transported by barge over water. Now, one technical problem the ancient merchant marine had to solve was getting heavy crates onto a boat. In those days, you see, boats were comparatively small, and when a massive cargo was carried across the gunwales, the boat turned turtle. So the traditional merchantman's crane was developed, using counterweights to maintain an even keel while loading. In the two hundred years since power winches became universal, it has been forgotten that today's basic machinery was developed when all the power had to be supplied by muscle. Winches in the confined area of a top deck are limited to small loads, and pulleys are limited by friction, unless fitted with roller bearings and chain. So when the ancient longshoremen had a heavy idol to ship to Thebes, they used counterweights to lift the thing as well as get it over the gunwale without capsizing the boat. Herodotus says that the Egyptians were exceedingly clever in the invention of counterweights.

Once the system of counterweights is developed, construcion practice inevitably turns to megaliths, because it is far easier to handle one big stone than to shape the hundreds of little blocks each megalith replaces. By far the biggest advantage of megalithic construction is permanence. Brick buildings can be torn down by every marauding horde that passes through the country, but no one is going to knock over a temple held together by fifty-ton lintels.

The counterweighted cranes developed to pass cargo between ship and shore need not be limited to a single lift. In crowded shipyards today, sections of ships are transported across busy work areas through the air, being passed from one crane to another. Workmen become so skillful that the loads may never be set down between first lift and final deposit. The same principle is practicable for the transport of building stone from the prehistoric quarry to the job site. When heavy cargo is shipped out by the millions of units over hundreds of years, economics justifies the construction of a chain of cranes, set about fifty yards apart, to transport material in a continuous stream over all kinds of terrain—like a pipeline, the most efficient of all material transports.

The design of the merchant ship's crane is scaled up for overland transport of ancient building materials. The base of the transport crane is a stone disc, about six to eight feet in diameter and a foot or two thick. The bottom surface is flat enabling it to rest securely on the ground; the top surface is fashioned into a shallow cone. Another stone disc of the same dimensions is set onto the base. The bottom surface of the top disc is made into a negative cone so that the wheel can rotate, balanced on the base without sliding off.

At opposite sides of the top surface of the upper wheel, hemispherical depressions are carved, about a foot deep and wide enough to contain the fire-hardened butts of a pair of booms. The booms can be extended by splints to extend over one hundred feet upward and outward at a forty-five degree angle. They are held balanced in position by a line passed over their upper ends, from which a heavy container is suspended at each end. Now, if a light line is suspended from the center of the main cross line, a small amount of force will pull the line down; closing the straight angle of the cross line brings the ends of the booms together. As the booms come together, rolling without friction in the cups holding their butts, the top ends rise, lifting the containers. The practical amount of rise is no more than a foot or two before the changing load vectors multiply the tension of the lines to their breaking point. But within this narrow range of efficiency, one man can raise many tons, and the requirements of transport can be served as soon as the load is clear of the ground.

In operation, cargo is loaded onto one container, while the other container is counterbalanced with ballast. When the cargo is a replicated building block, the weight remains constant, so there is no delay in balancing it with a counterweight of known measure. The control line suspended from the cross line carries a counterweight, too, sufficient to balance the container units, so the operator riding the top bearing stone need pull down on the control line with little more than the force required to overcome inertia. The control line also serves to take up slack

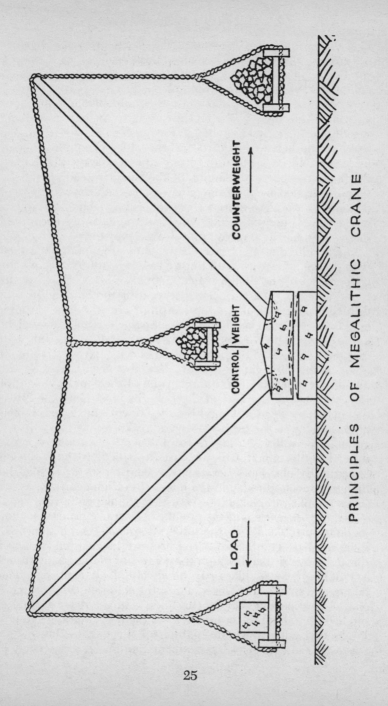

PRINCIPLES OF MEGALITHIC CRANE

COUNTERWEIGHT

CONTROL WEIGHT

LOAD

in the cross line as the hawsers stretch during service. Once off the ground, the load is carried around the reach of the booms by rotating the crane. The friction on the bearing stones is overcome by employing a draft animal.

A chain of these transport cranes would pass a continuous stream of building stone from the quarry to the city at walking speed, pausing only long enough to transfer the loads from one crane to another; while down the other side, an equal weight of ballast would be transported from city to quarry. The ballast side need not be deadheading all the time; instead of ballast, supplies needed by the quarry can be packed for transport in containers of unit weight. This system is low technology, labor intensive and non-polluting; when all the costs are accounted for, we can hardly improve on the efficiency today.

Once industrial machinery is put into production, the men who do the job always add improvements with or without the cooperation of the design office. Sooner or later, it will occur to a crane jockey that his pony is sweating too hard, so he will drill the center out of the top bearing to form a pipe through which he can pour clay slip to lubricate the bearing. The weight of the load will hold the bearing surfaces so tightly together, and they will be ground to such an even curve, that very little—if any—of the slip will be able to flow between them. So the operator will add a pole to the bottom of the counterweight of the control line; the butt of the pole fits nicely into the lubricating hole. When the weight is pulled down, the tamp is forced onto the clay in the pipe, and waddayaknow—pressure lubrication! An alternate means of lubrication probably used, instead of the fanciful design just described, is the inclusion of naturally oily shoes between the bearing surfaces, like the lignum vitae used today in the heaviest marine bearings.

A stationary crane has limited use, even in an industrial yard. When the ancient engineers designed the transport crane, they made it portable. When the job is completed, components are disassembled. The tamp for the pressure lubrication is passed through holes drilled through the centers of both bearing stones so that they will roll like a pair of wheels joined by an axle. The booms and spars are balanced across the axle. The whole bundle of sticks is tied together by the lines, and the draft animal is hitched up to haul the crane away to the next building site.

The first critic to whom I submitted this engineering system was a retired Master of Exams from Cambridge University, J.

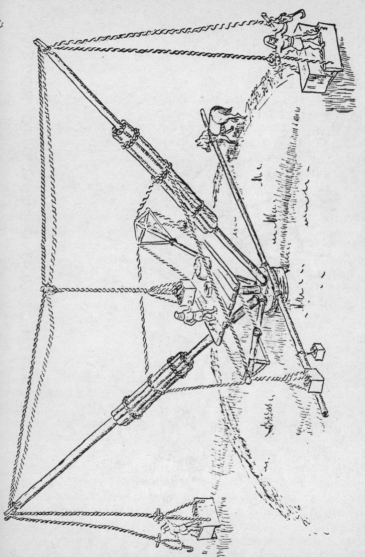

MEGALITHIC CRANE IN OPERATION

MEGALITHIC CRANE READY TO ROLL

Lloyd Brereton, to whom I am indebted for first publication. Brereton reasoned that transport cranes must have been as ubiquitous in the Age of Gold as trailer rigs are on today's highways, so he referred to the *Encyclopaedia Britannica* to see if any available evidence supports such speculation. Obviously, wood and rope do not outlast the pyramids, so we cannot expect to find any organic remains of the prehistoric transport system. But throughout the sands of the Middle East, archaeologists *have* found huge stone discs in considerable numbers. Tribespeople who found them first have been using these 'millstones of the gods' for thousands of years to grind their grain. The hole in the center of the upper disc is designed to pour in grist, and the conical bearing surfaces allow the flour to flow out around the edge as the stone is turned by a draft animal.

Naturally, modern scientists assume without question that the stones were manufactured for no other purpose. But it is curious that no other societies in the world have developed rotary millstones like the 'millstones of the gods'; everywhere else, grain is ground by women rubbing it back and forth in cylindrical bowls in mortar-and-pestle fashion.

The real mystery of the megalithic monuments is that no one over the past two thousand years was able to figure out a way to build them. Yet any competent tradesman could have told the authorities how to build these megalithic monuments, if anyone bothered to ask him. But no one *did* ask, because most people believe their authorities know everything. In Rhodesia there is a magnificent city built of precision masonry, constructed on a monumental scale comparable to the minor pyramids; the natives call the ruins Zimbabwe, and they attribute the construction to their ancestors. The European invaders found it necessary to establish the myth of white supremacy in order to justify their rape of the country. If the black people are capable of constructing cities like Zimbabwe, then the whites cannot possibly be biologically superior. So scientific authorities assured the world that Zimbabwe was constructed by a mysterious civilization, most likely the same people who built the pyramids; but whoever they were, they were not black. Now that Rhodesia has been taken over by democratic parties, the country is being called Zimbabwe by its own people, and the world is accepting as fact that the natives' ancestors are the people who built the monumental city.

*　　*　　*

This essay is not intended to explain how the prehistoric megalithic engineers *did* build their monumental temples. That question can be answered only by archaeologists on the basis of the evidence they dig up. The problem is, *how could it be possible* for a Neolithic society to have engineered the monuments we find today?

Now just because the engineering outlined in this essay is derived through the oldest traditions of the construction trade, and proven practical for megalithic monuments, it does not mean that the ancients used these methods or anything like them. Construction contracting is the most creative enterprise. At any given time and place, a contractor avails himself—from a variable selection of resources within his competence—to implement the technology he believes most likely to be executed profitably. A big-time operator like General Pyramids Consolidated, needing to keep its army of slaves busy earning its garlic during a slack century, may very well have put them to work piling earthen ramps up the sides of the pyramids with wicker baskets. On the other hand, I believe the way the ancients transported megaliths for their monuments was to attach a small tuning fork to each stone, causing the module to levitate when the properly tuned vibration was sounded. When a monolith is set to resounding, its vibrations keep it in the air most of the time. During the greater part of the wave cycle, when the mass is floating, a light touch will move it in any direction.

I believe ancient engineers used this technology because the ancient myths describe it. This application of musical theory is far too sophisticated for a Stone Age people to incorporate in their myths unless the authors actually witnessed the technology in operation.

You didn't think I was going to let you go without *some* speculation, did you?

2 This Crystal Planet

How to Create a Worldwide Communications Network—Still Using Bronze Age Technology

One disturbing puzzle of archaeological science is the number of similarities between some aboriginal cultures widely dispersed around the world. Pyramids, for example, are found not only in Egypt, but also in Peru, in Mexico, on the South Sea Islands, and in China. Moreover, many of the details of construction are identical. The present-day Indians of Peru build reed rafts identical to the ones pictured on the walls of ancient Egypt, and the modern Indian word for this craft is the same word spoken by the Nilots thousands of years ago. There are so many similarities between the Aztec tongue and classic Greek that the early European explorers believed they had discovered a lost Greek colony. Aztec-Greek lexicons have been compiled, and modern exploration is finding remains in South America indicating a Phoenician origin. In *Mysteries From Forgotten Worlds*, Charles Berlitz reproduces an 'alphabet' of pictograms found in the ruins of a prehistoric city of western India and compares it with the undeciphered writing of Easter Island; the glyphs are identical.

The evidence indicates a cultural migration from the centers of Mediterranean civilization across the Atlantic and around the world. Thor Heyerdahl proved this theory feasible by mounting a series of daring expeditions to cross the world's oceans on primitive rafts. Anyone who missed the films can read the adventures of this courageous amateur in his books, *Kon-Tiki* and *The Ra Expeditions*.

Heyerdahl proved that the Queen of Sheba could have hocked the crown jewels to outfit a task force of three papyrus rafts for crossing the Atlantic and finding a sea route to Cipango, dispersing Nilotic culture and smallpox en route. But no original thinker constructs a perfect and complete theory. Eventually so many loose ends accumulate that a better theory is required. Heyerdahl left an awful lot of loose ends.

Ever since the first sailor lost sight of land, survivors of shipwrecks have been cast up on faraway shores. But none of them transplanted a civilization. Aside from knowing how to build boats and sail them, most sailors were barely civilized enough to keep themselves stoned on grog. Any of them lucky enough to survive a greeting from headhunters went native as soon as they saw *wahine*.

Drifting sailors did not disperse culture; they dissipated it. It took centuries for Europe's highly organized and heavily supported American colonies to become self-supporting, and most of them could not survive from Thanksgiving to Christmas without help from the native Indians. For Mediterranean culture to have dispersed on papyrus rafts, the sea lanes between Memphis and Mexico must have carried more traffic than the great ocean liners during the heyday of American immigration.

One other problem Heyerdahl neglected to mention: Like wine and oysters, language does not travel well at pedestrian speeds. Within a generation, language changes; within a lifetime, classical Latin became Italian, French, Spanish, Rumanian, and Romansh. If a language is to maintain its identity around the world, rapid communication is required. If Nilotic civilization dispersed to the Americas and the South Pacific, it could not have drifted on rafts. Moreover, when the pyramid builders decided to disperse, they did not roam randomly, but proceeded *directly* to locations comprising a global network.

Inspection of a global relief map reveals that the eastern contours of the Americas conform closely enough to the western contours of Europe and Africa for the two sides of the Atlantic to have been joined together at one time. You will also notice a string of salt lakes, progressively diminishing in size, stretching from the Straits of Gibraltar through the Black Sea and along the northern base of the Himalayas, suggesting that an ocean passage once connected the Atlantic with the Pacific between Eurasia on the north and the Afro-Indian land masses on the south. People have noticed this structure since the first world maps were drawn and the Theory of the Continental Drift was proposed for serious scientific consideration over a hundred years ago. The original theory was disregarded until the undersea drilling of oceanographers after World War II brought up cores that could be explained in no other way.

As you can see for yourself, all the southern land masses

duplicate the same shape and contours. South America is a larger, elongated version of the Indian subcontinent. Africa is Australia in capitals. In addition, all these large blocks have a large island off their southeastern coasts: Tierra del Fuego, Madagascar, SriLanka, and Tasmania. Such regularity suggests that the continental slabs did not break off and drift in an entirely haphazard manner. The pattern suggests that the entire southern continent of Gondwanaland must have been formed as a regular crystal during the same heat when the geological crust solidified. The lines of cleavage not only follow regular fracture planes, but are also determined by a regular structure governing the dynamic tectonic forces that cause the Continental Drift. There may be a corresponding regularity in the original contours of the northern continents, but the outlines have been obliterated by the collisions that have occurred since. In artistic analysis, a general rule is to begin with the largest forms first, and then follow the natural subdivisions. The largest geographical feature on our planet is the Pacific Ocean. Notice that the Pacific Rim forms a rough circle defining about a third of the global surface. With such a large fraction of the geography covered by a circular basin, it follows that the center of the Earth's land masses must be approximately opposite the center of the Pacific. When the globe is turned around, you see that antipodal to the center of the world ocean is the region where Europe, Asia, and Africa are in continental collision.

The distribution of land and sea has all the appearance of a global flow carrying the continental crusts from a pole at the center of the Pacific to the opposite pole. The Americas and the World Island have been separated by a counterflow that does not extend into the circle of the Pacific Rim defining a third of the Earth's surface.

In order that Gondwanaland be formed around the South Pole, we may infer that another tectonic current existed during a primordial geological era, flowing south from the Equator, if not from the North Pole. This flow was reversed from south to north during an intermediary era before the present flow pattern became established. The fact that the South Polar region is a circular plateau ten thousand feet above the sea, whereas the North Polar region is a circular ocean basin, indicates that the tectonic current arises roughly around the South Polar Circle and stops roughly around the North Polar Circle. This coincidence, in turn, suggests the angle of the Polar Axis relative to the plane of the ecliptic is involved with the direction of the tectonic currents, just as this angle determines the currents of global weather patterns. In this era the tectonic currents appear to be aligned along an east-west axis. This suggests catastrophic shifts in the Polar Axis. Now, as it happens, drill cores brought up from the ocean beds do show sudden reversals of the Earth's magnetic field in geological times.

Because almost all the visible features of the planet are in the Eastern Hemisphere, a look at the center of the world may give some clue to account for the peculiar geological distribution. As soon as you perceive that the Pacific Rim draws a rough circle around the Nile Delta, you will notice that many of the lesser geographical structures—including undersea ridges—also form circles around the Delta. If you take a compass and scribe these circles, you will eventually have enough data to show that the major land forms follow arcs at intervals of 15 degrees from the Delta, especially the arcs at 90 and 120 degrees. These are the major intervals of spherical surface harmonic overtones.

The pattern is quite obvious; the Earth's surface is defined by vibratory currents. The Moon was found to ring like a bell when struck by the first Lunar Lander in a seismic test. If the Moon is a planetary crystal, why not the Earth?

The Great Pyramid at Giza is located at the center of the world's land masses and in the highest concentration of natural voltage combined with a tolerable climate. When the British inventor Sir W. Siemens climbed to the summit of the Great Pyramid, he found his body discharging sparks as if he were standing on a high-voltage coil. In fact, the atmospheric voltage on the Giza has been measured at 500 per vertical meter, and Siemens should be credited for discovering what Pyramid Power is all about.

The goal of Heyerdahl's first expedition was Easter Island (also known for manifesting very high electromagnetic potentials), the farthest east of a group of South Pacific islands where the pyramids of another prehistoric civilization have been found. And like the Egyptians, the aboriginal people of this region called themselves Children of the Sun. The South Pacific is at the antipodes—180 degrees from the Nile Delta.

In America, from the Great City of the Sun at prehistoric Cuzco to the Great Pyramid of the Sun at Teotihuacán, a third great-pyramid civilization arose. This culture ranged along an arc 120 degrees from Giza.

A quarter of the way around the world, 90 degrees in the other direction from Giza, recent archaeological digs have uncovered the greatest pyramids known—a thousand feet on a side —buried in the jungles of southern China, and thought to be natural mountains until the overgrowth was cut away. These mountains are the ancestral homelands of the Japanese, who still believe they are Children of the Sun.

The meaning of this peculiar dispersion was revealed by accident during the early days of World War II. At that time, radar spotters picked up the echoes of distant aircraft by listening through giant earphones instead of watching blips on fluorescent screens, as is done today. The crews frequently reported hearing whistles of no known cause. Eventually, research reported in *Scientific American* found an ionized layer in the upper atmosphere that filters radio waves selectively in the 7½-cycle-per-second band, and apparently reflects them back to Earth, bringing them to a focus at the antipodes.

When lightning strikes, a broad band of radio waves is emitted, to be heard as static in unfiltered radio sets. The 7½-cycle-per-second band is filtered from lightning discharging at the other side of the world and brought by the Schumann Layer to a focus 180 degrees distant, where the radar spotters heard the filtered static as whistles.

To military engineers, the 7½-cycle-per-second frequency possesses two properties of great value. The Schumann Layer guides this wavelength all around the world without losing signal strength, like light between two mirrors. This wavelength also penetrates water. The United States Navy realized that this was the frequency needed to keep in constant communication with its nuclear submarine fleet, constantly ranging beneath the surface of the world's oceans, and so it constructed a worldwide radio network broadcasting on the 7½ Hz band, called the Sanguine Project—later extended for civilian shipping as the Omega Navigation System.

The striking property of the 7½-cycle-per-second radio wave is that it is exactly 25,000 miles long. A wave broadcast at this frequency will expand in a growing circle until it is as long around as the equator, and then contract until it comes to a focus at the antipodes. At the other side of the world, it reverses phase and direction, expanding again to girdle the globe before returning to another focus at its exact point of origin. The loop of this wave arrives at its point of origin exactly in time to coincide with its own following wave. This means that only a single wave

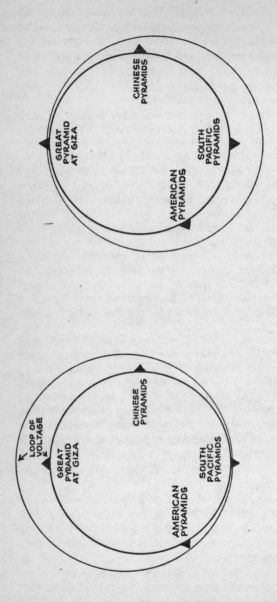

THE FUNDAMENTAL ELECTROMAGNETIC RESONANCE BETWEEN THE EARTH AND ITS SURROUNDING IONOSPHERE CREATES A VOLTAGE LOOP OSCILLATING FROM ONE SIDE OF THE WORLD TO THE OTHER, AS SHOWN IN THIS DIAGRAM. THE GREAT PYRAMID AT GIZA AND THE GREAT PYRAMIDS OF THE SOUTH PACIFIC ARE LOCATED WHERE THE NATURAL GLOBAL VOLTAGE ESTABLISHES THE HIGHEST ELECTRICAL POTENTIAL.

exists at any time, and that the wave encompasses the whole world. The entire planet beats electromagnetically at this frequency, like a cosmic heart. The 7½-cycle-per-second frequency is the fundamental period of surface resonance of the global crystal.

Resonance in the Schumann Layer means that overtones will be generated at one half the fundamental frequency, one third the fundamental frequency, and one fourth the fundamental frequency, and so on, in successive fractions. The length of the second harmonic overtone is one half the diameter of the Earth, the third harmonic overtone is one third the diameter of the Earth, and the fourth harmonic is one fourth the diameter of the Earth. The pyramid civilization of the South Pacific is halfway around the world from the Nile Delta, the pyramid civilization of America is a third of the way around the world from the Nile Delta, and the forgotten pyramid civilization of prehistoric China is a quarter of the way around the world from the Nile. The great pyramids of the world are located precisely at the places where the strongest electrical potential is concentrated on the Earth, and at the loops of the harmonic standing waves surrounding the Earth. The Schumann Layer functions as a radio loudspeaker with its diaphragm enclosing the entire planet.

The great pyramids of antiquity are a virtual duplication of the modern military radio-communications network, broadcasting on the 7½ Hz band from the main transmission tower on the Giza Plateau, with studios and executive offices in beautiful downtown Memphis.

If the laws governing radio engineering are correctly stated, the pyramids of the world must function as a planetary radio-communications network. The only question is whether the pyramid architects knew what they were doing when they built the original solid-state electronic modules. Ten thousand years from now, astro-archaeologists finding the Telesat Communications modules in stationary orbit around the Earth will likely ask if these electronic capsules were put into orbit (in locations corresponding to the Great Pyramid Network) entirely by accident. Or were they deliberately intended to function as a global communications system? In *Mysteries of the Great Pyramid*, Peter Tompkins reports that all the great cities of prehistory were built up around a central pyramid, and each was located at an exact degree interval from the Great Pyramid at Giza. Now, when settlers are founding a city, their prime

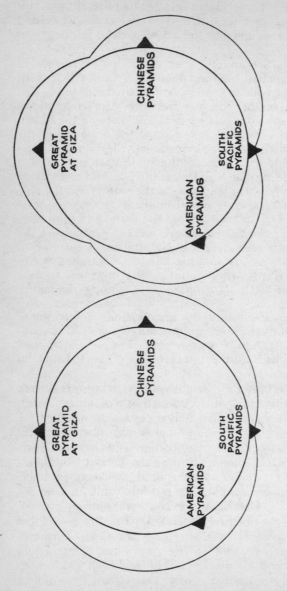

AS THE VOLTAGE LOOP OSCILLATES BETWEEN THE NILE DELTA AND THE SOUTH PACIFIC ISLAND AT THE ANTIPODES, AN EQUATORIAL LOOP IS ESTABLISHED BETWEEN THE TWO POLES.

THE CHINESE PYRAMIDS ARE FOUND WHERE THE EQUATORIAL LOOP ESTABLISHES THE HIGHEST ELECTRICAL POTENTIAL.

THE AMERICAN PYRAMID CIVILIZATION IS FOUND WHERE THE LOOP OF THE THIRD HARMONIC ESTABLISHES THE HIGHEST ELECTRICAL POTENTIAL.

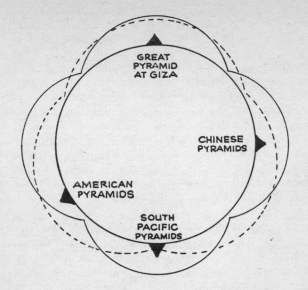

WHEN THE FUNDAMENTAL RESONANCE IS SUPERIMPOSED
WITH THE SECONDARY, TERTIARY AND QUATERNARY
HARMONIC, A STANDING-WAVE VOLTAGE STRUCTURE IS
ESTABLISHED, SURROUNDING THE EARTH WITH THE
ELECTRICAL VOLTAGE PATTERN SHOWN IN THIS DIAGRAM.
ALTHOUGH THE ELECTROMAGNETIC WAVES FLOW
AROUND THE WORLD, THE STANDING-WAVE PATTERN
REMAINS STATIC.

considerations are the availability of food, water, defense, and
trade. Why would a location at a precise degree interval from
the Great Pyramid on the Nile Delta be more important? Well,
no civilized city can endure without communications.

The radio diaphragm surrounding the Earth is powerful
enough to broadcast radio signals at least as far out into interplanetary space as Jupiter. Is there anyone out there listening?
Clear NASA photographs received from the space probes reveal
a complex of pyramidal structures on the Moon, called the
Blair Pinnacles. Had the same photographs been taken anywhere on earth, the Blair Pinnacles would likely be the ruins of
a prehistoric temple complex. Mariner 9 has sent back an
ambiguous photograph of a colossal pyramidal structure on
Mars. If the site proves* artificial, we may be living through

*Tom Bearden just reported new data supporting the likelihood of its
artificial nature.

39

the scene from *2001: A Space Odyssey* in which secrecy surrounds the discovery of a polished black monolith in the crater Clavius.

It is technologically feasible for a galactic civilization to implement the principles of pyramidal harmonics to construct a communications network throughout the Solar System, with relays carrying signals to the stars, requiring no power other than the natural resonance of organized space. Such a civilization could very well employ the natives for sweat labor just as the Pentagon hires Eskimos to clear the building sites for the DEW Line. As long as the aborigines can be induced to follow instructions, it isn't necessary that they have the slightest understanding of what they are doing.

Proof that the great pyramids of the world are solid-state electronic modules in a worldwide power-generating and broadcasting network was established over fifty years ago by that superhuman genius, Nikola Tesla. Tesla knew there is a powerful voltage between the Earth and the upper atmosphere; this is what causes lightning to discharge from the Earth to the sky. When an antenna is raised, this voltage climbs it, in a gradient to be concentrated at the tip. Any fluctuation in the natural voltage between Earth and sky generates a flow of electricity in the antenna. This is how a radio receiver works.

In radio's early days, the current generated in the receiving antenna was fed into a resonating circuit that would filter out the wavelength of the broadcasting station to which it was tuned and amplify the signal until it had enough energy to actuate earphones.

Tesla figured that the Earth would filter out its resonant frequency, amplify the current by its resonance, and function as a capacitor in a circuit as big as the entire planet. He pumped an electric current into the Earth, and tuned to a precise harmonic length of the Earth's resonant frequency. As he expected, the electric wave traveled to the antipodes and bounced back in time to coincide with its own following wave, doubling its amplitude. After repeated amplification in this manner, the electric energy burst from the top of Tesla's pyramidal tower to illuminate the entire countryside with the most spectacular artificial lightning storm ever seen. The surge melted the wiring in all the power generators serving the whole county. No one would supply Tesla with electricity after that.

Undaunted by superabundant success, Tesla continued his

experiments to prove that when his antenna was pumping electric waves into the Earth, a standing-wave pattern was generated in the planet's geo-electromagnetic structure. At precise degree intervals, or harmonic fractions thereof, all Tesla had to do was drive a metal rod into the ground and plug in the electric frying pan, hair dryer, or washing machine. Tesla was the last of the big-time pyramid architects.

It was Nikola Tesla who electrified the world with alternating current. To the end of his life, Tesla dreamed of recycling all those unsightly high-voltage electric power transmission towers and replacing them with broadcast power generated and received through the Earth. But if power could be tapped from the Earth, how would Consolidated Edison-General Electric-Standard Oil bill you for it?

If we can all communicate on Pyramid Power, where does Ma Bell cut in? When he died in New York City on a winter night in 1943, Tesla was alone in a hotel room, possessing little more than when he had arrived as an immigrant with four cents in his pockets. The Secret Service immediately sealed his room, and whatever papers he had were transferred to government vaults, where they remain to this day.

There is one more discovery to be made by an inspection of a global map. The range of the great-pyramid civilizations is limited to the Tropic Zone, but is not quite parallel to the equator. The Great Pyramid at Giza is just about as far north of the Tropic of Cancer as Easter Island is south of the Tropic of Capricorn. If we tilt the Tropic Zone to pass through those two extreme points, we find that the realigned Tropics contain the Pyramid Zone neatly between them.

If the Tropics are tilted in this way, the new North Pole is located in Greenland—which brings up a curious coincidence. Immanuel Velikovsky states that the North Pole was situated in Greenland during an earlier geophysical period. Look at your globe and draw a circle around the area covered by ice during the last era of planetary glaciation. You will find that its center is in Greenland. Now, how it is possible for Siberia to foster subtropical flora and fauna while Chicago is buried under miles of ice, unless the North Pole is in Greenland?

This suggests that the Great Pyramid Broadcasting System depends upon receiving solar radiation at 90 degrees for the most efficient operation. The source of energy must have come from cosmic space. The power generated by the pyramids

41

depends upon the passage of the Sun, the positions of the planets, and the altitude of the ecliptic. The most important duty of the Priests of the Sun was the forecasting of the ephemerides, and—following the construction of the Great Pyramid—their supreme technological achievement was the perfection of mathematical instruments for celestial observation.

The proclamation that pyramid mathematics was developed to forecast the annual flooding of the Nile could be made only by scholars who never worked on a farm. Farmers don't need such minute precision; for thousands of years, crops have been harvested despite religious calendars—the Moslem calendar, for example—running as much as half a year out of sync with the seasons.

The original network was likely constructed before the last geological catastrophe which shifted the Polar Axis. The original pyramid could be counted on to generate vibrations for millennia without maintenance, except for adjustments to the calendar from time to time. The power stations could even be left as unmanned posts checked occasionally by servicemen. Resident technicians needed no more knowledge of how the solid-state modules worked than the billing office of the Tennessee Valley Authority knows about electrical wiring. Over thousands of years of undisturbed operation, the regular duties of the staff would degenerate into superstitious ritual (just as occurs in every public-service bureaucracy). A major disturbance of the Earth's electromagnetic parameters, however, would result in an immediate, complete, worldwide power failure. And no one would know what to do.

Sudden and violent change in the Earth's electromagnetic structure includes a breakdown of the radiation shield in the ionosphere, admitting massive floods of cosmic radiation to the Earth's surface. All vital and radioactive processes would be affected, so that the carbon dating process archaeologists rely on may not be reliable for times prior to such a catastrophe. (Naturally, professional archaeologists resist any evidence that the carbon dating on which all their work is based may not be valid.) The Great Pyramid at Giza may be very much more ancient than archaeologists are wont to believe; all the other lesser pyramids may very well be comparatively recent copies.

In some civilized centers, the civil service was well enough established to construct more pyramids to replace the ones which no longer functioned at design efficiency, and the dispersed tribes built smaller pyramids where they settled. These copies were built on a smaller, human scale, however, instead of the former superhuman proportions; the massive megaliths were reduced to adobe bricks. But since the builders had lost the science of harmonics (if they ever possessed it in the first place), they copied their pyramids precisely in all the wrong places. When these new structures failed to function, the ignorant and terrified people sacrificed everything they had to the Sun, imploring it to restore function and power to the pyramid. The Aztecs commemorated these supplications and sacrifices until the modern era.

As long as the American physicist remains in his lab and the American archaeologist remains in his dig, neither specialist will ever suspect that each sees but a part of the whole truth.

Although the Soviet system is inefficient in delivering consumer goods to its people, Communists are more receptive to revolutionary engineering. So while American authorities deny Tesla's discoveries, Soviet engineers have built a complex of Tesla towers in Karelia. Independent amateurs in America have found that the Soviets are using them to alter the natural pattern of electromagnetic standing waves surrounding the Earth. Because the atmosphere carries electrical charge, air currents tend to follow geo-electromagnetic flow patterns. When the natural standing-wave pattern is changed, the climate changes accordingly. Have you noticed?

Independent researchers, comparing the locations of the exceptional number of severe earthquakes during the past decade with the report of Soviet nuclear tests, conjecture that the Reds are using seismic waves to plot the pattern of standing waves defining the interior of the Earth: They give the planet a concentrated shock at a fault line and wait to see where the impulse emerges on the other side of the world. As long as local authorities deny the existence of this technology, how can the victims prove their earthquake was not a natural event?

One of the most competent authorities on Soviet military technology is Tom Bearden, described by *Fate* magazine as a one-man think-tank who overawes his colleagues by his high-octane genius. After retiring from the United States Army with a rank of lieutenant colonel, Bearden was employed by the military defense establishment to design war games, program computers to play them, and serve as chief planning officer

43

for interpreting the results in deploying America's nuclear missile system.

Bearden has come to the conclusion that the sonic booms and flashes of light over the northeastern American sea-board during 1976–77 were the effects of plasma vortices the Soviets' Karelian Tesla towers have succeeded in transmitting through the Earth.

In theory, Bearden says that these plasma vortices can pack the zap of a hundred-megaton hydrogen bomb. But not to worry; the Soviets are merely target practicing with low-caliber ammunition. Completely retired now, Bearden is organizing amateur scientists throughout the world to establish evidence that the Soviets are really doing these things. (The findings are published in a quarterly bulletin available from P.O. Box 1182, Huntsville, Ala. 35807.) Curiously, his international organization has recently been decimated by a spate of illness, death, violent attacks, destruction of papers and property, and official harassment—everyone who makes noises about Tesla's inventions receives the same treatment. Bearden is playing General Billy Mitchell; if no one listens, it may be Pearl Harbor all over again.

3 Beyond Velikovsky

Einstein's Relativity Demonstrated, Mining Pure Energy from Empty Space, and the Green Hills of Mars

Scientific theory preaches that our Solar System condensed from a cloud of galactic dust. Mutual gravitational attraction is calculated to have drawn the nebulous dust particles toward a common center, inducing rotational velocity in the process. When density of dust at the center became high enough, pressure generated heat for the nuclear fusion of a star to develop. Centrifugal force flattened out the primordial nebula into a disc, and then streamers were thrown off to condense independently, forming a system of planets. Photographs of nebulas in deep space illustrate each stage of the evolutionary process, so why couldn't it have happened here just like that?

Well, to begin with, just about every planet in the Solar System exhibits some anomaly that proves the Nebular Hypothesis could not possibly have happened the way we are told in school.

If you refer to the figures supplied by Funk and Wagnall Encyclopedia, you will see that Jupiter has one tenth the diameter of the Sun, but it rotates a hundred times faster. Volume is diameter cubed, and momentum is one half of velocity squared. Therefore, the Sun has a thousand times the volume of Jupiter, but Jupiter has five times the angular momentum. Jupiter is also a fairly dense planet, containing more mass than all the rest of the Solar System combined, outside of the Sun. Therefore, even after we make allowances for difference of radius, quick and dirty calculation indicates that Jupiter has about as much angular momentum as all the rest of the Solar System put together, including the Sun.

But that should not be! You see, if all the rotational energy in the Solar System is generated by gravitational collapse, then almost all the momentum in the Solar System must be concentrated in the Sun. While not altogether impossible, it is exceedingly unlikely for any planetary material to be flung away from a condensing core without losing the rotational momentum it acquired by condensing in the first place.

Venus rotates backward on its axis, and radiates more energy than it receives from the Sun. At some stretch of the definition, Venus can be regarded as a miniature star. Dust clouds whirling in space are almost certain to be as homogeneous as any other gaseous mixture, so why is the chemistry of Venus grossly different from all the other inner planets? And why did Venus just happen to lock into a three-two orbital-rotational ratio with the Earth without inducing corresponding changes in the angular momentum of our planet? We are, after all, more alike in size and mass than any other bodies in the system, so any gravitational friction between us should be settled by mutual compromise; instead, Venus gave in to us entirely.

The Earth has by far the largest satellite in the Solar System, and the farthest away. The moon, in fact, is not a satellite of the Earth at all. Moon and Earth form a double planetary system, both going around the Sun while the Moon is perturbed to cross and recross the Earth's orbit. If you draw the orbits of the Earth and the Moon to scale, you will see for yourself that the Moon never curves around the Earth as the orbit of a true satellite must. The Moon always falls in a curve to the Sun—sometimes more, sometimes less. When the curve is tighter, the Moon accelerates and passes the Earth; when the curve is less, the Moon falls back until it is swept back across the Earth's orbit again.

The polar axes of the Earth and Mars are tilted nearly 30 degrees from the plane of their orbits. When a nebula condenses into a disc, all spins must be in the same plane. One of Mars's satellites, Phobos, completes an orbital revolution before Mars makes a polar rotation. Such excess of angular velocity makes Phobos one body in the solar system that could have been cast off by centrifugal force, but it is generally agreed to be a captured asteroid.

The Nebular Hypothesis does not account for the swarms of broken rocks and planetoids orbiting the sun between Mars and Jupiter.

Jupiter radiates more energy than it receives from the Sun, and seems to be growing hotter. Jupiter really *is* a small star. If so, then the Jovian satellites will eventually be the ideal homes for life to evolve in the Solar System, independently of the

Earth. They already have an atmosphere and ice; they need only heat.

Saturn is lighter than water, and stands out with its well-developed ring system. Faint rings have been found circling Jupiter and Uranus too, so it may be supposed that all giant planets have the gravitational intensity sufficient to hold interplanetary dust. The Nebular Hypothesis does not explain why Saturn should be so egregiously favored while Jupiter has practically no rings to show for itself.

The axis of Uranus and the revolutionary plane of all its satellites are at right angles to the plane of the rest of the solar system. As Uranus makes its annual rounds of the Sun, Uranians see the Sun spiral out from one pole until it circles the equator and then spiral in toward the opposite pole, then back again. While the Sun is over the equatorial region, day and night resemble our daily cycle on Earth, but during the season when the Sun shines directly over a pole, it is seen to circle around the horizon of the respective temperate region giving constant daylight for (Earth) years without interruption; the equator receives constant twilight while the Sun is over either pole. It is impossible for the velocities of the Uranian system to have evolved at right angles to the solar nebula that created it; it had to have been rolled over onto its side by an interplanetary collision.

The amount of energy required to rotate the axis of a spinning body 90 degrees is equal to its rotational momentum. Uranus is one of the giant planets; an impact sufficient to knock a giant planet onto its side could be delivered by nothing less than a direct impact with another giant planet or a near collision with a celestial body many times more massive. But no other planet can possibly exist in the same region of Uranus' solar orbit while traveling in the opposite direction, necessary if a collision is to take place. Therefore, the anomaly of Uranus can be accounted for most satisfactorily by proposing that an alien star entered the Solar System on a collision course from the interstellar reaches.

If the alien star were the same size as Uranus, a direct impact would be necessary to make the Uranian system turn right about. But a *direct* impact would have exploded both celestial bodies and nothing would exist in the Uranian orbit today except another ring of interplanetary rubble. On the other hand, an alien star the size of the Sun would have torn the entire Solar System to pieces by its massive gravitational field. Therefore, we can infer that the alien must have been about ten times the mass of Uranus, and probably traveling at a speed in the range of solar escape velocity.

Neptune has lost control of its satellites. The inner one revolves backward as a singular retrograde planetoid, while the outer satellite has the most eccentric orbit in the entire Solar System, not including the comets.

Because of its small size—and an orbit so eccentric that its radial differential is greater than the entire Solar System, inclusive of Neptune—it is generally accepted that Pluto must have been a satellite of Neptune that escaped into solar orbit. When professional astronomers admit this much and no more, they are as guilty of prevarication as ecologists who do not mention that oil is burned to charge the batteries of electric cars. You see, there is nothing at the outermost planetary orbit of the Solar System with sufficient energy to multiply the orbital velocity of Pluto by a factor of $\sqrt{2}$ in order that it can escape Neptune.

Unless you postulate an adventitious interplanetary collision course arriving from interstellar space! Every planet in the Solar System violates the physical requirements of the Nebular Hypothesis. An alien star is the obvious solution to a host of astronomical embarrassments. But it also blows a hole through the theory of planetary evolution. This is exactly what Immanuel Velikovsky did.

When Velikovsky published *Worlds in Collision*, he became the victim of most vehement and scurrilous persecution. To all of us plebes who tread streets of concrete, it doesn't make a damn bit of difference how the Solar System came to be in the shape it is in during our mayfly lives. One story is as good as another as long as it stops the kids from asking 'How come?' when you want to put them to bed.

This *story* is not a scholarly examination of Velikovsky's theory to determine its validity. But if you look at some evidence everyone can see for himself with a toy telescope, it appears that authorized science is what remains after the editors have left all the contradictory evidence on the cutting-room floor. Let's run the astronomical scraps through the old Moviola to see if we can splice something together that will satisfy the popular demand for sex and violence without overlooking that the Solar System is as scrambled as an air terminal after a bomb blast.

As we fade in, Star Rangers scanning the heavens from their far-flung outposts on top of prehistoric ziggurats sight a 'thing' from outer space on a collision course with the Solar System. Of course in prehistoric times, climbing the steps of a two-hundred-foot pyramid was the service range of the contemporary astronauts, but it doesn't really matter whether an expeditionary force intercepts the 'thing' while it is still zillions of parsecs distant at Warp 7, or whether we merely wait until it comes within the Three-Mile Limit and reveals itself. I mean, unless you can evacuate the entire planet, there is nothing you can do about a cosmic 'thing' anyway! You can bet your Trekkie button the Federation isn't going to tell the taxpayers the options have expired before everyone can plainly see for himself that it isn't pie in the sky. It is a matter of common knowledge that several asteroids have come close to the Earth in recent years, and who cared, as long as they stayed out of sight?

So when we cut in the next scene, we see that the 'thing' is already in the Solar System, close enough for the ancients to see that it is a small dark star. As it passes by Neptune, at the Federation frontier, it is seen on the video scope to reach Neptune's outermost satellite, Triton, just past conjunction. The advent of massive gravitational attraction causes Triton to stop in orbit and plummet into its giant primary. As it falls, Triton gains enough velocity to reestablish itself in another orbit, but revolving in the opposite direction, as it would when negatively accelerated. This is where we find Triton today.

The middle satellite is approaching conjunction with the dark star, so it is accelerated positively until it leaves Neptune and goes into the planet business by itself as Pluto. And this, O Best Beloved, is why we find Pluto returning to its original orbit around Neptune every year, to this very day, just as it must according to the law by Newton. The innermost satellite, Nereid, is in opposition, so the stellar fly-past merely perturbs it into extreme eccentricity.

Velocity in interstellar space is low on the cosmic speedometer, so the dark star is drawn into solar orbit by the close encounter with Neptune.

As it continues its wild fall toward the Sun, the dark star passes through Uranian territory. The titanic waves raised by the interplay of gravitational and electromagnetic fields draw long streamers of gas and particles from the alien and rocks the Uranian System onto its side. The star is now a comet, the most stupendous comet ever seen by mankind, with its dragon head containing more mass than the entire Solar System exclusive of the Sun, and a fiery dragon tail drawn halfway round the sky, illuminating all the heavens, day and night.

The most concentrated mass of the tail whirls around itself by the snap of the gravitational whip until it condenses into a ball surrounded by rings of the lighter trailing particles. And this, O Best Beloved, is why Saturn is a giant planet lighter than water, as if it were entirely atmospheric, and it is surrounded by a system of rings today. Vestiges of the cometary tail remain yet as rings surrounding Jupiter and Uranus.

Careening drunkenly now, the dark star strikes the next planet, shattering it to bits. The explosion throws the alien into a stable solar orbit and it takes up permanent residence in our system as the violent overlord, Jupiter. And this, O Best Beloved, is why we find most of the angular momentum in the solar system today to be in the Jovian System; it was the ante the dark star brought with it from the interstellar reaches so that it could sit in on our game. We lost the planet Krypton, but the Earth is saved in the final reel. The debris of the interplanetary explosion fills the orbit of Sol 5 with asteroids, a ring around the Sun like a petrified rainbow signifying God's favor with man. Lest you leave this theater thinking there is no scientific proof for this scenario, Dr. Bill Ovendin, a professor at the University of British Columbia, recently programmed a computer with the orbits of all the known comets and discovered that they trace back to a single explosive point in the asteroid belt.

For human interest between the special-effects shots in interplanetary space, we segue Earthside, and pick up the legends overlooked by Velikovsky. In some early era Neptune and Uranus were visible. They were the original titans, and it was natural that Uranus, being the nearest and brightest, would be regarded as the creator of the heavens.

Saturn was called Kronos by the ancient Greeks. Kronos became Chronos, the God of Time. If Saturn has ever occupied the orbit now holding Jupiter, Saturn's original period would be twelve years, and its location in the zodiac would tell the houses of the horoscope. The planet would serve as a celestial time-keeper for astrologers, a profession enjoying the highest social status in those days. As the new cosmic timekeeper, it was natural for Saturn to usurp the liege formerly bestowed upon Uranus; imperial jealousies are never generously disposed, so it was natural for the Titans to be buried in the black of deep space. Neptune being put out deeper than Uranus, Poseidon became god of the oceans.

But *kronos* is also the etymological root of corona, meaning *crown*. Encircled by shining rings, Saturn is indeed the crowned planet, deserving recognition as the supreme god. Now, the rings of Saturn are barely visible to the sharpest eyes; a simple magnifying mirror will bring them into view. If, however, Saturn ever sat on the throne of the Jovian orbit during an interim reign while the crown prince was off to the wars with Krypton, the crown of Kronos would be clearly visible throughout the Solar System. The Saturnian crown is still used artistically to represent the halo of angels, so we may suspect that men once saw the King's Father directly.

Saturn is said to have eaten his children, fatally overlooking Jupiter. This part of the Greek myth could be describing knots of the original cometary tail condensing into visible satellites before spiraling into their primary as the comet settled into stable orbits. The surviving rings we see today may be merely the last vestige of a multijeweled crown.

The myths indicate that there were several changes in the planetary orbits during the creation of the heavens as we now see them. The various poetic histories conform to the hypothesis of a dark star entering the Solar System because it is unlikely for all planets to be aligned so that everything happened during the course of a single preorbital trajectory. Once caught in a cometary orbit after being trapped on Neptune's gravitational trident, Jupiter probably made hundreds of elliptical circuits between successive encounters of the catastrophic kind, leaving plenty of breathing space for mankind to recover between cosmic disasters and make up different myths.

Here is a synopsis of Immanuel Velikovsky's scenario: When Jupiter collided with Krypton, a mass was blown out denser than the Saturnian comet, including some of the rocky core of the doomed planet. The jetsam formed a new, smaller comet swinging tightly around the Sun on an orbit of fifty-four years. Because space is tight in the downtown section of the solar community (hold tight while we change metaphors again), this comet had close and frequent encounters with all the inner planets. On each strafing run, the inner planets were peppered with meteors. The craters are clearly visible today on Mercury, the Moon, and Mars. The terrestrial craters have been obliterated by our fertile atmosphere, but the fracturing of the Canadian Shield in concentric circles around Hudson's Bay indicates that the northern sea is a meteor crater created by an

51

impact powerful enough to shake the Earth on its axis and probably obliterate large life forms over half a hemisphere; smaller craters pock the entire Atlantic Coast of North America. After a pyrotechnic exchange with Mars, which must have rivaled the spectacle of Jupiter and Uranus for mankind, the daughter comet swung into stable orbit around the Sun as the planet Venus.

The persecution of Velikovsky is most determined from astronomers because it destroys the steady-state theory necessary to make coherent sense out of current measurements. It is necessary to believe that all extant physical activity occurs everywhere at all times at the same rate in order to have a theoretical science and get paid for teaching it. Otherwise all you have is engineering and science fiction. The physical evidence on which Velikovsky has been prosecuted by the establishment is not proclaimed to the public; the excuse is that the public does not understand physics. The result is that Velikovsky was given no chance to defend himself in a manner that habeas corpus is supposed to provide, nor could he put judgment to a public jury.

Insofar as I have been able to get any answer from the enemies of Velikovsky, they base their case on one of Newton's laws conserving momentum. A material body disturbed in its orbit must return to the point of disturbance on each subsequent revolution. The Solar System preserves far too much order today for any collision courses to have occurred in the past, much less the historical past. Most physicists, who are specialists, are satisfied that the law is inviolate. But Einstein opened a completely new and radical conception of Newton's laws.

The solution to Velikovsky's problem is implicated with the solution to another irritating problem bothering astronomers for over two hundred years, Bode's Law of Planetary Harmonics.

Since it became possible to calculate the distance separating planets with more than accidental precision, it has been known that each planetary orbit is twice as far from the Sun as the one nearer; credit for measuring the intervals has been accorded to the German astronomer Johann Elert Bode, for getting his ratios recognized by publication in a hard-core scientific journal. Bode found that if an Astronomical Unit is established as 9,000,000 miles, the following multiples of that Unit are measured between the planets: Sun 4 Mercury 3 Venus 3 Earth 6 Mars 12 Asteroids 24 Jupiter 48 Saturn 96 Uranus 96 Neptune 96 Pluto.

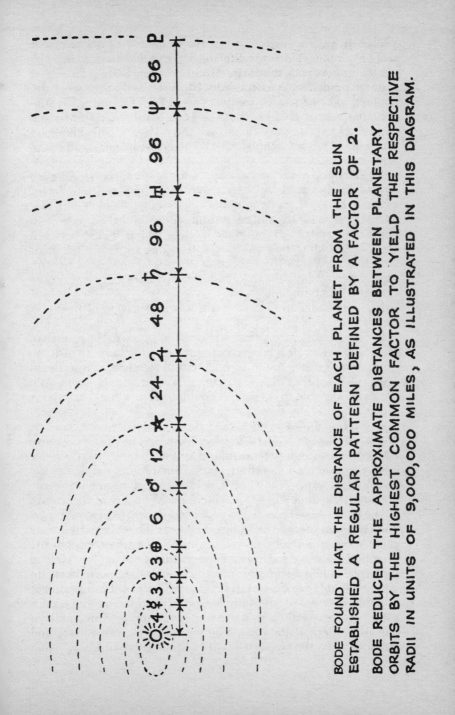

BODE FOUND THAT THE DISTANCE OF EACH PLANET FROM THE SUN
ESTABLISHED A REGULAR PATTERN DEFINED BY A FACTOR OF 2.

BODE REDUCED THE APPROXIMATE DISTANCES BETWEEN PLANETARY
ORBITS BY THE HIGHEST COMMON FACTOR TO YIELD THE RESPECTIVE
RADII IN UNITS OF 9,000,000 MILES, AS ILLUSTRATED IN THIS DIAGRAM.

Saturn was the outer limit of the known system when Bode did his number, so the discovery of Uranus was accompanied by great professional excitement to see that Bode's Law was followed indefinitely. But when Neptune and Pluto were discovered to be separated by intervals of 96 Units each, the professional verdict decided that the Law had broken down; and respectable orthodoxy regards the Theory of Planetary Harmonics as an empirical accident with no significance whatsoever.

Upon immediate inspection, the interval between each planet, from Venus to Uranus, increases by a constant factor of two. Bode's Law would be indisputable were it not that Mercury is twice the distance from Venus that the progression demands, and Neptune divides the orbital space between Uranus and Pluto by two. Two is the constant factor throughout, but why does the operation change at the extremes? An almost perfect law like this is what keeps scientists awake more than the 2 A.M. nursing.

Now, in order to understand the conformation of the Solar System, it is exceedingly difficult to avoid the theory of cosmic collisions. In order to understand what happens in the process of cosmic collisions, it is necessary to prove Bode's Law. In order to prove Bode's Law, you must understand relativity. Professional scientists insist that no one can understand relativity—not even Einstein was able to do that. The impossible will not take a bit longer. Let me show you.

If you find a thin, vibrant disc, dust the surface with powder, and make it resonate to a musical tone, the powder will move into concentric windrows, spaced a wavelength apart. As the circles approach the center, radial vibrations break up the circular pattern. What you see is a model representing the mechanics of Bode's Law and the General Theory of Relativity.

What you see happening is produced by waves of sound traveling through the disc in all directions. As the waves move, they carry the dust with them. Where two waves meet in opposite directions, a wave pattern is created that does not travel. The pattern is called a standing wave. The standing wave beats up and down in the middle of its length, while there is no motion at all on both ends. As an example, two children swinging a skipping rope between them are making a standing-wave pattern with the length of the rope. The wave energy travels back and forth between the two arms swinging the rope; if the traveling

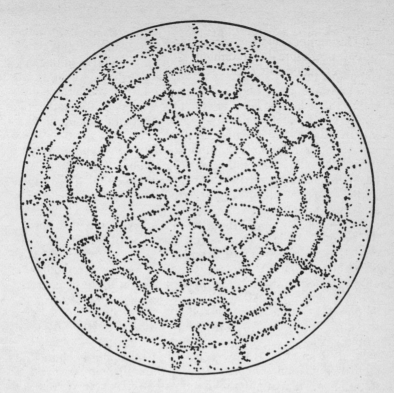

DISTRIBUTION PATTERN MADE BY POWDER ON A
CIRCULAR VIBRATING PLATE

waves coming from both ends are not synchronized, the standing-wave pattern breaks down so that you can see the wave motion travel from one end to the other and back again. A ring strung on the rope will be driven by the waves from the loop in the middle to the nodes at the ends.

Most of us armchair physicists will argue at great length before being stirred to dust a disc with powder or observe children skipping rope. But if you will ring for your maid to bring you tea, you will see concentric standing waves on the surface of the fluid in the cup when she puts the tray down on your end table with sufficient impact to generate vibrations.

Those of you who study any one of several versions of this model will eventually notice that the rings representing planetary orbits around the Sun are spaced equal distances apart, whereas the planetary orbits have a difference between each orbit established by the factor of two. The reason the rings of powder on the disc are equally spaced is that the vibration is supplied by waves of equal length. The field of energy constituting the Solar System is defined by waves of all lengths, like white sound. When a full-frequency spectrum of sound is used to generate standing waves around a focal point the longer waves are superimposed through the shorter waves amplifying them in some places and canceling them out in others. The harmonic beats where all waves coincide to amplify and cancel in the same places form rings separated by octave intervals, just as we see in the Solar System.

Since Niels Bohr published the planetary model of the atom, the atom and the Solar System have been understood to be models of each other. Since Erwin Schrödinger proved the wave equations of quantum physics, the atom has been recognized as a standing-wave structure composed of an infinite spectrum of random vibrations. All we have done is say that if the Solar System is a cosmic atom, and if the atom is a standing-wave structure, then the Solar System must be a standing-wave structure too.

Well, if an atom is the essential unit of material, and if an atom is a standing-wave structure, then all material must be standing-wave structures. Therefore, if the solar system is a standing-wave structure, then the entire gravitational field of the Sun must be solid *material*. The only difference between the material of the Sun's gravitational field and the Earth we stand on is density. The wavelengths of gravity are so long that we can walk through them as if the standing waves were insubstantial, like we walk through air. Many people will snort that this deduction is preposterous; the more schooling you have, the more likely you will insist that it isn't so. But have you forgotten that material, by definition, is identical to mass? Einstein's General Theory of Relativity proves that the Sun's gravitational field possesses mass and functions as a massive structure. Einstein knew it all along; he just stopped talking at this point.

Now, why do you suppose that the curtain of silence is dropped at the point where Einstein and Bode and Velikovsky all meet with mutually supporting calculations?

Well, when Einstein published his equation $E = mc^2$, he proved that all material can be converted into energy, the amount of energy depending upon the wavelength of the standing waves forming the material. This means that the Sun's gravitational field—filling all of interplanetary space—is *solid energy*, just like uranium, waiting to be mined. Now you can see why Einstein is supposed to be incomprehensible, and relativity is taught to make sure no one can understand it. And now you can see why Bode's Law cannot be allowed to be recognized. And now you can understand one reason why Velikovsky had to be given the deStalinization treatment. You see, the military-industrial-financial establishment can stake out a monopoly on uranium, enabling them to charge whatever OPEC demands for oil, but everyone has free access to the Sun's gravitational field. As Al Capp illustrated with his shmoos, industrial society cannot survive if the universe supplies mankind with the necessities of life free of charge.

The military-industrial establishment published the first release of atomic energy from uranium in 1945. Would you like to guess when Einstein's equations were proven by releasing the energy contained in the Sun's gravitational field? According to information available to me, Nikola Tesla gained insight into spacic energy at the beginning of this century, before petroleum was used for much except kerosene lamps; by 1925 he was ready with experimental proofs. Reports have reached me of about a dozen private parties making similar discoveries in their own way; they were all given the deep six, like Tesla got. The latest word is that a party in the Northeast, calling itself X-Tec, is trying to get into the consumer market by generating spacic energy from the Sun's gravitational field; we'll see if it survives any longer than its predecessors. In the meantime, the Soviet military-industrial establishment has managed to secure a monopoly on the free air their citizens are allowed to breathe, so it has no reservations about implementing Tesla's theories for generating electric power from the energy of empty space, and they are proceeding apace while Americans are building windmills to get the same energy the hard way.

The most completely documented chronicles of the development of spacic energy in America is reported in the book, *The Sea of Energy*, published privately by the son of the inventor in Salt Lake City. As far back as 1914, before the First World War, Henry Moray, an electrical engineer, perceived that solar activity generated radio waves in the Earth's atmosphere. An antenna transforms the electromagnetic waves into a standing electrical wave in the conductor. The energy of this wave is amplified by resonance to provide the power for the primitive radio sets Moray played with. From the operation of the crystal radio, Henry Moray acquired an understanding of the Sun's gravitational field as a cosmic standing-wave structure supported by incredibly high-frequency vibrations from the quantum field.

Natural germanium crystals were used for tuning the early radio receivers. Moray inferred that there was something in crystal geometry that could resonate in tune with the quantum field and transform the field energy into electricity by harmonic amplification. So Moray sought the purest germanium crystals chemists could provide, always complaining that the crystallization was not fine enough. Once the alternating current in his crystals was fed into a step-down transformer, Moray drew enough power to supply household appliances with 500 kilowatts indefinitely.

As a reward for bringing his invention to the United States Government, Moray became the target of hired assassins. His laboratory was destroyed and his credibility ruined so that the design of his 'valve' was lost with his death. We can infer, however, that the essential design must be so simple that you can pick up all the components to build your own basement electric-power generator from Radio Shack for less than one thousand dollars. After all, in those days not even Moray had enough knowledge of electronics to build a transistor, and the complexity of a TV circuit was beyond conception. However he did it, he tapped infinite energy, like the technologists of Atlantis, without pollution or explosion from the Sun's gravitational field with a circuit essentially no more complicated than a glorified crystal radio receiver.

Now that we have made the engineering discoveries which are the goal of this story, we can return along the trail on which we came to prove that the evidence for the prosecution of Immanuel Velikovsky by the defenders of the established faith is unconstitutional. At the same time, we shall make further discoveries opening new possibilities for more fantastic engineering.

We see the powder bounce up and down on the vibrating disc until it settles at the static nodes of the standing waves. Obviously there is a tangible flow from the loops to the nodes. Therefore, any particle supported by the vibrating field will flow from the loops to the nodes like a chip on a stream of water. The loops can be represented as hills and the nodes as valleys for purposes of illustration, but we can see there are no real hills or valleys in a gravitationally defined space; there is only a flow of energy represented by velocity gradients. So any celestial body let loose in the solar gravitational field will 'gravitate' toward the nodal orbits as if attracted by some mysterious force. You will recognize that this model is the very representation of Einstein's General Theory of Relativity.

Now you know why a planet finds an orbit at a specific location in the solar gravitational field, but you may wonder why it revolves around the Sun in that orbit. If you examine the vibrating plate intently, you will eventually discern that the flow of energy does not stop at the nodes, but undergoes an abrupt right-angle rotation of velocity with a transformation of wave frequency and speed. The flow now circles the orbital path, carrying the planet with it like a log on a stream of water between banks.

A scientific intelligence can be gauged by the point at which a person accepts an answer as final and satisfying. So we continue to ask why a planet occupies a specific location in its orbit instead of any other location along the length of the circuit. Well, more inspection eventually reveals that the entire nodal circle vibrates as a linear standing-wave structure, composed of all frequencies, like a circular violin string strummed by a breeze. The fundamental frequency of the orbital resonance is subdivided into harmonic intervals, defined by loops and nodes of different energy levels. Residual discords in the harmonics cause the pattern to revolve around the field center. Therefore, once a body is carried by the solar gravitational field to an orbital node, it will be carried farther in the orbit until it settles into the deepest node in the circuit. That is where it will stay as the entire gravitational field revolves.

If this model is true, then there must be a node of the second gravitational harmonic of the Earth's orbit directly opposite us, on the other side of the Sun where we can never see it. This calculation is the probable source of the myth of a counter-Earth. The myth, in fact, is true, but its proof is another story for another book. As far as we are concerned here, astronomers have proven that the counter-Earth cannot possibly exist because there is no evidence of the gravitational effects its mass must produce. The reason the counter-Earth is not found where

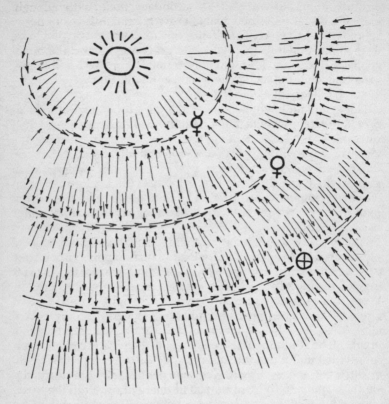

UNIVERSAL ENERGY IN THE QUANTUM FIELD FLOWING
PERPENDICULAR TO THE PLANE OF THE ECLIPTIC MEETS
IN MUTUAL OPPOSITION IN THE INTERORBITAL SPACES.
DIRECT OPPOSITION ROTATES THE FLOW BY 90° AND REDUCES
ITS VELOCITY BY A FACTOR OF C, TRANSFORMING THE QUANTUM
FIELD INTO THE SOLAR ELECTROMAGNETIC FIELD.

THE ELECTROMAGNETIC FLOW IS RADIAL ON THE PLANE OF THE
ECLIPTIC. WHEN THE OPPOSING CURRENTS MEET, ANOTHER
ROTATION OF 90° REDUCES WAVE VELOCITY BY A FACTOR
OF C, AGAIN, TRANSFORMING THE ENERGY INTO THE
GRAVITATIONAL ORBITS CARRYING THE PLANETS ON ITS STREAM.

we see it is that the space in which it is calculated to exist happens to be where the fundamental loop also exists. The fundamental loop cancels the secondary node with enough energy left over to raise a sizable gravitational 'hill.'

But the fundamental loop flows in *both* directions around the orbit to the tertiary nodes, a pair of gravitational nodes located 120 degrees in both directions from us. Therefore we should expect to find smaller masses of interplanetary material caught in these two depressions. In fact, astronomers have observed aggregations of interplanetary debris collected at both these locations. When we see interplanetary bodies attracted to locations where no central masses exist to establish a gravitational field, we have the proof that gravity is inherent in the flow of energy through space rather than a mysterious force generated by a massive body.

We are forced to the further deduction that Jupiter does not possess the most powerful gravitational field in the Solar System because it possesses almost all of the mass, outside of the Sun. It is possible for Jupiter to have been *created* the most massive body in the Solar System because the greatest gravitational flow in the Sun's field is into that location. Jupiter's gravitational field would exist at that location even if Jupiter were not there to occupy the space.

Another proof for the model is derived from the empirical fact that all planets sweep an equal area of their orbital planes in equal times. Momentum is a time function, and the equation for momentum is identical to the equation for circular area. Therefore, the angular momentum of each orbit is identical throughout the solar gravitational field. The energy flow from the loops to the nodes is in direct proportion to the area of the orbits. The area of the orbits increases at a ratio equal to half the square of the radius. First sight tells you that the momentum of each orbit should increase by one half the square of the radius if all the energy from the loop area flows into the nodal line. But the solar field is a compound standing-wave structure; the wavelength of the standing waves increases in direct ratio with the radius. The formula equating energy with wavelength states that momentum varies *inversely* with frequency, so that the difference in energy represented by area is exactly balanced by the difference in energy represented by wavelength. Therefore, the momentum of the solar field must be constant throughout.

A final proof is that escape velocity is $\sqrt{2}$ times orbital velocity.

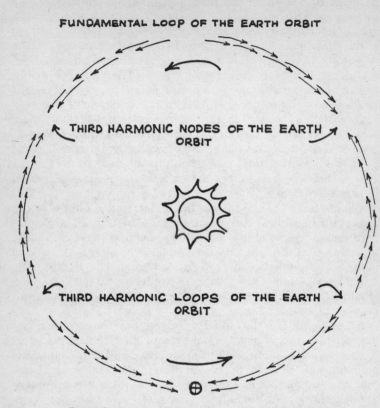

FUNDAMENTAL LOOP OF THE EARTH ORBIT

THIRD HARMONIC NODES OF THE EARTH ORBIT

THIRD HARMONIC LOOPS OF THE EARTH ORBIT

FUNDAMENTAL NODE OF THE EARTH ORBIT

THE CURRENT OF GRAVITATIONAL ENERGY FORMING THE
EARTH'S ORBIT RESONATES LIKE A MUSICAL STRING WOUND
IN A CIRCLE AND JOINED BY A KNOT. THE EARTH
GRAVITATES TO THE PRIME NODE. SMALLER MASSES
OF INTERPLANETARY FLOTSAM HAVE GRAVITATED TO THE
NODES OF THE TERTIARY HARMONIC OVERTONE. LOOPS AND
NODES OF HIGHER HARMONIC OVERTONES SUBDIVIDE BOTH
THE ORBITAL AND INTERORBITAL SPACE INTO A CONSTANTLY
CHANGING PATTERN OF "HILLS" AND "HOLLOWS" REVOLVING
AROUND THE SUN AT ORBITAL VELOCITY.

Because the increase in field wavelength is continuous along the gravitational radius, the loop between successive orbits is not exactly between them. The peak of the loop divides the inter-orbital radius into fractions defined by $\sqrt{2}$, like the intensity of light falls off from a point source. An increase of orbital velocity by a factor of $\sqrt{2}$ gives a planet exactly the amount of momentum it needs to surmount the gravitational hill confining it to its orbital valley; and once over the divide, it is downhill all the way to the next star.

The planets occupy the deepest nodes in the solar gravitational field. But the solar space is filled with an infinite number of higher harmonic nodes, each of which is capable of attracting and holding a mass of material. But the higher the harmonic, the less energy flows into it, so the smaller the mass it can attract and retain.

The gravitational node holding the Moon, for example, is so much weaker than the one holding the Earth that by the time a spaceship acquires enough velocity to escape the Earth, it has far too much momentum for the Moon to hold. You may think that landing on the Moon must be as easy as falling out of bed once an Apollo spaceship passes over the gravitational divide and begins to drop down onto the Moon. But this isn't true. Unless the space shot is aimed for a direct hit on the surface of the Moon, anything leaving the Earth must fly right past the Moon into solar orbit. Getting down is the hardest part.

Residual discords in the solar geometry causes the gravitational hollows to wander along courses described by the Drunkard's Walk, calculated by the mathematics of Probability. The amount of indirection is directly proportional to the value of the harmonic fraction. This means that the major harmonics represented by the planetary orbits will manifest negligible perturbations during any historic period, while the locations of higher harmonic nodes wander around with increasing randomness. Harmonic coincidence of wandering nodes results in periodic intersections; and once again, the lower the harmonic, the longer the period between intersections, while the higher harmonics are mixing together continuously. A small body caught in the shallow depression of a high node will be tipped out when confluence with a harmonic loop causes a temporary reversal of gravitational flow. If a particle is small enough, it can be carried by waves from node to node throughout interstellar space before it gets trapped by the powerful energy flow of a low harmonic. In fact, interstellar space is filled with dust floating around like this. Only the confluence of a low-harmonic loop with a low-harmonic node is sufficient to reverse the gravitational flow keeping a planet in its orbit. When that happens, planets, stars, and even galaxies are set on collision courses. Human history would be briefer than our two thousand years if low-harmonic loops intersected with low-harmonic nodes more often.

Now we can prove why Venus maintains a near-circular orbit today even though it is possible for her to have arrived in our skies within human memory after a series of catastrophic interplanetary collisions. Celestial bodies of planetary mass gravitate quickly to low-harmonic nodes, and low-harmonic nodes are inherently stable in orbit. Lighter celestial bodies can be retained more or less indefinitely in a succession of higher-harmonic nodes; the higher the harmonic, the more eccentric and unstable the orbit will be. This is why we find the planets to follow regular orbits of circular dimensions while the most eccentric and unstable bodies in the Solar System are also the lightest and most vaporous—the comets.

The flow of energy along high-harmonic orbits, not concentric with the field's center, are probably the 'rivers of energy' calculated by Professor Eric Laithwaite, which he hoped to harness with his gyroscopes. Once these currents are charted, they can be used to carry vehicles throughout the universe like floating on a cosmic jet stream across oceans of space and even time. But this is not a prevenient discovery, either. The Orientals have always known of the existence of Ley Lines, and they use Ley Lines to control the flow of energy in their social and personal space.

Meanwhile, back at the Kangaroo Courthouse, let's see how Velikovsky is making out with his defense. If he uses the proof of Bode's Law to disallow the invocation of Newton's law, Velikovsky runs the risk of incriminating himself by failing to foresee all the inevitable consequences of General Relativity. You see, all the octave orbits are occupied in the solar subdivision. If Venus settled on the residential lot at #2 Sunshine Valley, she probably had to evict a previous tenant. Whom do you suppose got bumped? When Venus arrived, the Moon moved in with Mother, and she has been spending the nights with us ever since. Plato mentions a time when there was no moon in the sky, called the Preselenite Era.

As soon as you begin to play musical orbits with the planets, the distribution of angular momentum in the Solar System is disturbed. When Venus brought additional energy from interstellar space, it meant that the gravitational hollow at #3 Sunnydale Circle must have drifted away from the Sun, while the revolution of the Moon made further adjustments to maintain Conservation. If the proposition of interplanetary collision is essentially valid, the Earth must have been closer to the sun during a previous geophysical era. Merely a few megamillion miles nearer would have added enough extra insolation to provide the Earth with an allover suntan from pole to pole. Geologists know that the Earth *did* have a tropical climate during the Paleozoic Age. Various hypotheses have been proposed to explain this anomaly, but they all require special cases involving a great deal of critical ignorance to remain tenable. Special cases do not a science make; the theory of interplanetary collision eliminates all the obvious anomalies with a single, simple, and provable mechanism.

Now if the Earth were forced to leave the Sunshine Coast of the Solar Estates and rebuild in the cooler hinterlands, whom do you suppose *we* displaced. It had to be Mars. If Mars were ever closer to the Sun than it is right now, it must have enjoyed the most invigorating climate in the entire Solar System while dinosaurs were sweltering in the swamps of the Tethys Sea. If life is a natural and inevitable process of planetary evolution, intelligent life had to appear on Mars before it began anywhere else.

Mars is large enough to hold an atmosphere capable of supporting terrestrial life forms. There is incontrovertible evidence of large flows of water on Mars. Where you find heat and light and soil and water, you must find life, too, or else you must throw out the Theory of Human Evolution from natural causes and return to Divine Creation.

The professionally published explanation for the evidence of liquid water flowing on Mars is that volcanic heat produced massive eruptions of water from sub-Arean rocks which streamed in a deluge for thousands of miles, carving canyons dozens of miles deep in the Martian continent. The professional statement may, in fact, be true, but the authorities cannot expect any high-school kid to believe it unless a lot of obvious contradictions are accounted for.

To begin with, it is almost as cold as liquid air on Mars; that is carbon dioxide we see frozen at the polar caps through our toy telescopes. Water is not going to flow very far from the source of eruption unless the entire hemisphere is warmed by volcanism to keep the water liquid. Enough heat to warm an entire hemisphere for swimming is not localized volcanism. It is either planetary convulsions or else it is climate.

Next, the range of temperature in which water can flow any distance over the land surface of Mars is restricted to the range of human comfort. Water cannot exist on Mars today because of the near-vacuum conditions, but even if Mars had all the air it could hold, the boiling temperature of water would still be so low that it would vaporize from the heat of volcanism before it flowed any distance.

Finally, the surface gravity of Mars is so low that water isn't going to flow rapidly enough to gouge canyons in its path unless it is constantly cascading. The canyons of Mars are not the kind of erosion carved by sheets of cascades, but more like the Grand Canyon, requiring a constant flow of water in the same channel for millions of years. This condition could not have been satisfied unless there was enough air on Mars to make a Tibetan feel at home, while the climate was no warmer than the USA and no colder than Canada. The absence of densely grouped meteor bombardment and other evidence of erosion subsequent to the carving of the Arean canyons indicates that Mars enjoyed a terrestrial climate very recently in astronomical time—probably within human memory. Mars probably lost its air, water, and climate by a sudden catastrophe involving heat intense enough to flash its oceans into water vapor.

If Velikovsky's scenario were staged during human memory, the people who saw the action were not living on Earth. They were Martians. As Mars is older than the Earth in its physical evolution, people on Mars may be expected to be more advanced technologically than we are now. Once they saw that alien squatters were going to break their leasehold on Sunny Acres, everyone who could raise the price of a ticket on the space ark would pack their luggage with Kodachrome, disconnect the household utilities, and hop over to the Mediterranean Riviera for the Ice Age—bringing their pyramid and crystal technology with them.

Support for the hypothesis of interplanetary colonization is found in comparative anatomy. The human animal is rather poorly adapted to conditions on Earth. The sunlight is too bright for our eyes and skin, our backbones are too long, our feet are

66

too weak, and our weight is ruinous to posture. If all creatures' morphology is determined by an evolutionary adaptation to its environment, we are best adapted for comfortable living on Mars, providing the climate were sweetened.

The white race appears to be new to this planet; civilization appears to have arrived suddenly and fully blown, then decaying for millennia after its initial appearance before recovering to our present eminence. The prehistoric aborigines of the cradles of civilization say they were conquered by Aryan invaders. The word *Aryan* is a variation of *Arean*, and *Arean* means *Martian*. There is a tribe in the Mideast who claims its ancestors came from Mars. As myths go, this one requires a sophistication of concept not proper to a primitive tribe; the Earth must be recognized as a planet unsupported in empty space, and the bright reddish star must be recognized as another world.

The most interesting evidence for the Martian origin of the white race, if not an extraterrestrial immigration for all of mankind, is provided by science reporter Gay Gaer Luce in *Body Time*. Biologists acknowledge that the vital rhythms of all creatures are established by their place of origin; this is why subequatorial plants bloom for us in the North at Christmas, and species entrained to the periods of the Moon are believed to have evolved in the tidal ocean. When experimental human subjects are isolated from all terrestrial rhythms as much as possible to discover the free period of the human body, it has been found that most people settle into a daily rhythm of sleep and work of twenty-four hours and forty minutes. Twenty-four hours and forty minutes is the daily period of Mars.

The mechanics of evolution make it practically impossible for any species to develop without integration with its environment, yet this is just what we have managed to do. We are profoundly out of order with the rest of terrestrial life. Everywhere we go, we destroy the natural environment with wholesale insensitivity to its rhythms and balances. We are reconstructing the entire surface of this planet to suit our alien dimensions. The distress we stir on this Earth may be due to the fact we don't belong here.

4 How To Build A Flying Saucer

After So Many Amateurs Have Failed

At the end of the nineteenth century, the most distinguished scientists and engineers declared that no known combination of materials and locomotion could be assembled into a practical flying machine. Fifty years later another generation of distinguished scientists and engineers declared that it was technologically infeasible for a rocket ship to reach the moon. Nevertheless, men were getting off the ground and out into space even while these words were uttered.

In the last half of the twentieth century, when technology is advancing faster than reports can reach the public, it is fashionable to hold the pronouncements of yesteryear's experts to ridicule. But there is something anomalous about the consistency with which eminent authorities fail to recognize technological advances even while they are being made. You must bear in mind that these men are not given to making public pronouncements in haste; their conclusions are reached after exhaustive calculations and proofs, and they are better informed about their subject than anyone else alive. But by and large, revolutionary advances in technology do not contribute to the advantage of established experts, so they tend to believe that the challenge cannot possibly be realized.

The UFO phenomenon is a perversity in the annals of revolutionary engineering. On the one hand, public authorities deny the existence of flying saucers and prove their existence to be impossible. This is just as we should expect from established experts. But on the other hand, people who *believe* that flying saucers exist have produced findings that only tend to prove UFOs are technologically infeasible by any known combination of materials and means of locomotion.

There is reason to suspect that the people who believe in the existence of UFOs do not want to discover the technology because it is not in the true believer's self-interest that a flying saucer be within the capability of human engineering. The true believer wants to believe that UFOs are of extraterrestrial origin because he is seeking some kind of relief from debt and taxes by an alliance with superhuman powers.

If anyone with mechanical ability really wanted to know how a saucer flies, he would study the testimonies to learn the flight characteristics of the craft, and then ask, 'How can we do this saucer thing?' This is probably what Wernher Von Braun said when he decided that it was in his self-interest to launch man into space: 'How can we get this bird off the ground, and keep it off?'

Well, what is a flying saucer? It is a disc-shaped craft about thirty feet in diameter with a dome in the center accommodating the crew and, presumably, the operating machinery. And it flies. So let us begin by building a disc-shaped air foil, mount the cockpit and the engine under a central canopy, and see if we can make it fly. As a matter of fact, during World War II the United States actually constructed a number of experimental aircraft conforming to these specifications, and photographs of the craft are published from time to time in popular magazines about science and flight. It is highly likely that some of the UFO reports before 1950 were sightings of these test flights. See how easy it is when you *want* to find answers to a mystery?

The mythical saucer also flies at incredible speeds. Well, the speeds believed possible depend upon the time and the place of the observer. As stated earlier, a hundred years ago, twenty-five miles per hour was legally prohibited in the belief that such terrific velocity would endanger human life. So replace the propellor of the experimental disc airfoil with a modern aerojet engine. Is Mach 3 fast enough for believers?

But the true saucer not only flies, it also hovers. You mean like a Hovercraft? One professional engineer translated Ezekiel's description of heavenly ships as a helicopter-cum-Hovercraft.

But what about the anomalous electromagnetic effects manifest in the space surrounding a flying saucer? Well, Nikola Tesla demonstrated a prototype of an electronic device that was eventually developed into the electron microscope, the television screen, and an aerospace engine called the Ion Drive. Since World War II, the engineering of the Ion Drive has been advanced as the most promising solution to the propulsion of interplanetary spaceships. The Drive operates by charging atomic particles and directing them with electromagnetic force as a jet to the rear, generating a forward thrust in reaction. The advantage of the Ion Drive over chemical rockets is that a spaceship can sweep in the ions it needs from its flight path, like an aerojet sucks in air through its engines. Therefore, the ship must carry only the fuel it needs to generate the power for its chargers; there is no need to carry dead weight in the form of rocket exhaust. There is another advantage to be derived from ion rocketry: The top speed of a reaction engine is limited by the ejection velocity of its exhaust. An ion jet is close to the speed of light. If space travel is ever to be practical, transport will have to achieve a large fraction of the speed of light.

In 1972 the French journal *Science et Avenir* reported Franco-American research into a method of ionizing the airstream flowing over wings to eliminate the sonic boom, a serious objection to the commercial success of the Concorde. Four years later a picture appeared in an American tabloid of a model aircraft representing the state of current development. The photograph shows a disc-shaped craft, but not so thin as a saucer; it looks more like a flying curling stone. In silent flight, the ionized air flowing around the craft glows as a proper UFO should. The last word comes from an engineering professor at the local university; he has begun the construction of an Ion-Drive Flying Saucer in his backyard.

To the true believer, the flying saucer has no jet. It seems to fly by some kind of antigravity. As antigravity is not known to exist in physical theory or experimental fact in popular science, the saucer is clearly alien and beyond human comprehension. But *antigravity* depends up on what you conceive *gravity* to be, doesn't it?

For all practical purposes, you do not have to understand what Newton and Einstein mean by gravity. Gravity is an acceleration downward, to the center of the earth. Therefore, antigravity is an acceleration upward. As far as practical engineering is concerned, any means to achieve a gain in altitude is an antigravity engine. An airplane is an antigravity engine, a balloon is an antigravity engine, a rocket is an antigravity engine, a stepladder is an antigravity engine. See how easy it is to invent an antigravity engine?

There are three basic kinds of locomotive engines. The primary principle is traction. The foot and the wheel are traction engines. The traction engine depends upon friction against a surrounding medium to generate movement, and locomotion can proceed only as far and as speedily as the surrounding friction will provide. The secondary principle is displacement. The balloon and the submarine rise by displacing a denser medium; they descend by displacing less than their weight. The tertiary drive is the rocket engine. A rocket is driven by reaction from the mass of material it ejects. Although a rocket is most efficient when not impeded by a surrounding medium, it must carry not only its fuel but also the mass it must eject. As a consequence, the rocket is impractical where powerful acceleration is required for extended drives. In chemical rocketry, ten minutes is a long burn for powered flight. What is needed for practical anti-gravity locomotion is a fourth principle which does not depend upon a surrounding medium or ejection of mass.

You must take notice that none of the principles of locomotion required any new discovery. They have all been around for thousands of years, and engineering only implemented the principle with increasing efficiency. A fourth principle of loco-motion has also been around for thousands of years: It is centri-fugal force. Centrifugal force is the principle of the military sling and medieval catapult.

Everyone knows that centrifugal force can overcome gravity. If directed upward, centrifugal force can be used to drive an antigravity engine. The problem engineers have been unable to solve is that centrifugal force is generated in all directions on the plane of the centrifuge. It won't provide locomotion unless the force can be concentrated in one direction. The solution of the sling, of releasing the wheeling at the instant the centrifugal force is directed along the ballistic trajectory, has all the ineffi-ciencies of a cannon. The difficulty of the problem is not real, however. There is a mental block preventing people from per-ceiving a centrifuge to be anything other than a flywheel.

A bicycle wheel is a flywheel. If you remove the rim and tire, leaving only the spokes sticking out from the hub, you still have a flywheel. In fact, spokes alone make a more efficient flywheel than the complete wheel; this is because momentum goes up only in proportion to mass but with the square of speed. Spokes are made of drawn steel with extreme tensile strength, so spokes alone can generate the highest levels of centrifugal force long after the rim and tire have disintegrated. But spokes alone still generate centrifugal force equally in all directions from the plane of rotation. All you have to do to concentrate centrifugal

71

force in one direction is remove all the spokes but one. That one spoke still functions as a flywheel, even though it is not a wheel any longer.

See how easy it is once you accept an attitude of solving one problem at a time as you come to it? You can even add a weight to the end of the spoke to increase the centrifugal force.

But our centrifuge still generates a centrifugal acceleration in all directions around the plane of rotation even though it doesn't generate acceleration equally in all directions at the same time. All we have managed to do is make the whole ball of wire wobble around the common center of mass between the axle and the free end of the spoke. To solve this problem, now that we have come to it, we need merely to accelerate the spoke through a few degrees of arc and then let it complete the cycle of revolution without power. As long as it is accelerated during the same arc at each cycle, the locomotive will lurch in one direction, albeit intermittently. But don't forget that the piston engine also drives intermittently. The regular centrifugal pulses can be evened out by mounting multiple centrifuges on the same axle so that a pulse from another flywheel takes over as soon as one pulse of power is past its arc.

The next problem facing us is that the momentum imparted to the centrifugal spoke carries it all around the cycle with little loss of velocity. The amount of concentrated centrifugal force carrying the engine in the desired direction is too low to be practical. Momentum is half the product of mass multiplied by velocity squared. Therefore, what we need is a spoke that has a tremendous velocity with minimal mass. They don't make spokes like that for bicycle wheels. A search through the engineers' catalog, however, turns up just the kind of centrifuge we need. An electron has no mass at rest (you cannot find a smaller minimum mass than that); all its mass is inherent in its velocity. So we build an electron raceway in the shape of a doughnut in which we can accelerate an electron to a speed close to that of light. As the speed of light is approached, the energy of acceleration is converted to a momentum approaching infinity. As it happens, an electron accelerator answering our need was developed by the University of California during the last years of World War II. It is called a betatron, and the doughnut is small enough to be carried comfortably in a man's hands.

We can visualize the operation of the Mark I from what is known about particle accelerators. To begin with, high-energy electrons ionize the air surrounding them. This causes the betatrons to glow like an annular neon tube.

Therefore, around the rim of the saucer a ring of lights will glow like a string of shining beads at night. The power required for flight will ionize enough of the surrounding atmosphere to short out all electrical wiring in the vicinity unless it is specially shielded. In theory, the top speed of the Mark I is close to the speed of light; in practice, there are many more problems to be solved before relativistic speeds can be approached.

The peculiar property of microwaves heating all material containing the water molecule means that any animal luckless enough to be nearby may be cooked from the inside out; vegetation will be scorched where a saucer lands; and rocks containing water of crystallization will be blasted. Every housewife with a microwave oven knows all this; only hard-headed scientists and softheaded true believers are completely dumbfounded. The UFOnauts would be cooked by their own engines, too, if they left the flight deck without shielding. This probably explains why a pair of UFOnauts, in a widely published photograph, wear reflective plastic jumpsuits. Mounting the betatrons outboard on a disc is an efficient way to get them away from the crew's compartment, and the plating of the hull shields the interior. At high accelerations, increasing amounts of power are transformed into radiation, making the centrifugal drive inefficient in strong gravitational fields. The most practical employment of this engineering is for large spacecraft, never intended to land. The flying saucers we see are very likely scouting craft sent from mother ships moored in orbit. For brief periods of operation, the heavy fuel consumption of the Mark I can be tolerated, along with radiation leakage—especially when the planet being scouted is not your own.

When you compare the known operating features of particle centrifuges with the eyewitness testimony, it is fairly evident that any expert claiming flying saucers to be utterly beyond any human explanation is not doing his homework, and he should be reexamined for his professional license.

For dramatic purpose, I have classified the development of the Flying Saucer through five stages:

Mark I—Electronic centrifuges mounted around a fixed disc, outboard.

Mark II—Electronic centrifuges mounted outboard around a rotating disc.

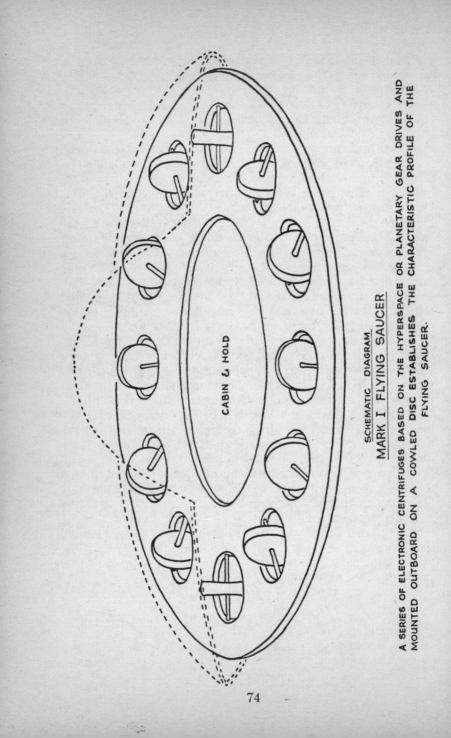

CABIN & HOLD

SCHEMATIC DIAGRAM
MARK I FLYING SAUCER

A SERIES OF ELECTRONIC CENTRIFUGES BASED ON THE HYPERSPACE OR PLANETARY GEAR DRIVES AND
MOUNTED OUTBOARD ON A COWLED DISC ESTABLISHES THE CHARACTERISTIC PROFILE OF THE
FLYING SAUCER.

Mark III—Electronic centrifuges mounted outboard around rotating disc, period of cycles tuned to harmonize with ley lines, for jet assist.

Mark IV—Particle centrifuge tuned to modify time coordinates by faster-than-light travel.

Mark V—No centrifuge. Solid state coils and crystal harmonics transforms ambient field directly for dematerialization and rematerialization at destinations in time and space.

Now that the UFO phenomenon has been demystified and reduced to human ken, we can proceed to prove the theory. If your resources are like those of the PLO, you can go ahead and build your own flying saucer without any further information from me, but I have nothing to work with except the junk I can find around the house.

I found an old electric motor that had burned out, but still had a few more turns left in it. I drilled a hole through the driving axle so that an eight-inch bar would slide freely through it. I mounted the motor on a chassis so that the sliding bar would rotate in an eccentric cam. In this way, one end of the bar was always extended in the same direction while the other was always pressed into the driving axle. As both ends had the same angular velocity at all times, the end extending out from the axle always had a higher angular momentum. This resulted in a concentration of centrifugal acceleration in one direction. When I plugged in the motor, the sight of my brainchild lurching ahead—unsteadily, but in a constant direction—gave me a bigger thrill than my baptism of sex—lasted longer, too. But not much longer. In less than twenty seconds the burned-out motor gasped its last and died in a puff of smoke; the test run was broadcast on radio microphone but the spectacle was lost without television. Because my prototype did not survive long enough to make a run in two directions, I had to declare the test inconclusive because of mechanical breakdown. So, what the hell, the Wright brothers didn't get far off the ground the first time they tried, either. Now that I know the critter will move, it is worthwhile to put a few bucks into a new motor, install a clutch, and gear the transmission down. One problem at a time is the way it goes.

A rectified centrifuge small enough to hold in one hand and powered by solar cells, based on my design, could be manufactured for about fifty dollars (depending on production run and competitive bids). Installed in Skylab, it would be sufficient to keep the craft in orbit indefinitely. A larger Hyperspace Drive (as I call this particular design) will provide a small but constant acceleration for interplanetary spacecraft that would accumulate practical velocities over runs of several days.

It is rumored that a gentleman by the name of Dean invented another kind of antigravity engine sometime during the past fifty years, but I have been unable to track down any more information except that its design consists of wheels within wheels. A gentleman in Florida, Hans Schnebel, sent me a description of a machine he built and tested that is probably similar in principle to the Dean Drive. Essentially, a large rotating disc has a smaller rotating disc on one side of the main driving axle. The two wheels are geared together so that a weight mounted on the rim of the smaller wheel is always at the

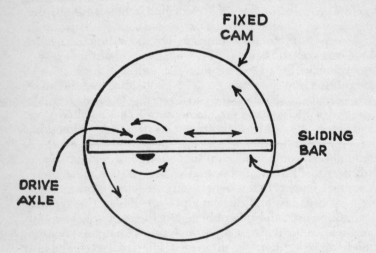

FIXED CAM

SLIDING BAR

DRIVE AXLE

ESSENTIAL DESIGN OF
THE HYPERSPACE DRIVE

THE ROTATIONAL VELOCITY OF THE OSCILLATOR
IS CONSTANT ON BOTH SIDES OF THE DRIVE
AXLE. THEREFORE, CENTRIFUGAL ACCELERATION
MUST ALWAYS BE HIGHER ON THE
LONG EXTENSION.

outside of the larger wheel during the same length of arc of each revolution, and always next to the main axle during the opposite arc. What happens is that the velocity of the weight is amplified by harmonic coincidence with the large rotor during one half of its period of revolution, and diminished during the other half cycle. This concentrates momentum in the same quarter continually, to rectify the centrifuge. The result is identical to my Hyperspace Drive, but it has the beauty of continuously rotating motion. Now, if the Dean Drive is made with a huge main rotor —like about thirty feet in diameter—there is enough room to mount a series of smaller wheels around the rim, set in gimbals for attitude control, and Mr. Dean has himself a Model T Flying Saucer requiring no license from the AEC.

In 1975 Professor Eric Laithwaite, Head of the Department of Electrical Engineering at the Imperial College of Science and Technology in London, England, invented another approach to harnessing the centrifugal force of a gyroscope to power an antigravity engine—well, he almost invented it, but he did not have the sense to hold onto success when he grasped it. Professor Laithwaite is world-renowned for his most creative solutions to the problems of magnetic-levitation-propulsion systems, and the fruit of his brain is operating today in Germany and Japan; his railway trains float in the air while traveling at over 300 miles per hour. If anyone can present the world with a proven and practical antigravity engine, it must be the professor.

Laithwaite satisfied himself that the precessional force causing a gyroscope to wobble has no reaction. This is a clear violation of Newton's Third Law of Motion *as generally conceived*. Laithwaite figured that if he could engage the precessional acceleration while the gyroscope wobbled in one direction and release the precession when it wobbled in other directions, he would be able to demonstrate to a forum of colleagues and critics at the college a rectified centrifuge that worked as a proper antigravity engine. His insight was sound but he did not work it out right. All he succeeded in demonstrating was a *separation between action and reaction*, and his engine did nothing but oscillate violently. Unfortunately, neither Laithwaite nor his critics were looking for a temporal separation between action and reaction, so the loophole he proved in Newton's Third Law was not noticed. Everyone was looking for action *without* reaction, so no one saw anything at all. Innumerable other inventors have constructed engines essentially identical to

Laithwaite's, including a young high-school dropout who lived across the street from me.

Another invention is described in U. S. Patent disclosure number 3,653,269, granted to Richard Foster, a retired chemical engineer in Louisiana. Foster mounted his gyroscopes around the rim of a large rotor disc, like a two-cylinder flying saucer. Every time the rotor turns a half cycle, the precessional twist of the gyros in reaction generates a powerful force. During the half cycle when Foster's gyros were twisting in the desired direction, his clutch grabbed and transmitted the power to the driving wheels. During the other half cycle, the gyros twisted freely. Foster claimed his machine traveled four miles per hour before it flew to pieces from centrifugal forces. After examining the patents, I agreed that it looked like it would work, and it certainly would fly to pieces because the bearing mounts were not nearly strong enough to contain the powerful twisting forces his machine generated. Foster's design, however, cannot be included among antigravity engines because it would not operate off the ground. He never claimed it would, and Foster always described his invention truthfully as nothing more than

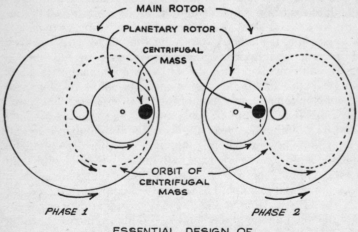

ESSENTIAL DESIGN OF
PLANETARY GEAR DRIVE

WHEN THE CENTRIFUGAL MASS IS AT THE RIM OF THE MAIN ROTOR,
THE VELOCITY OF THE MASS IS ADDED TO THE VELOCITY OF THE RIM.
WHEN THE CENTRIFUGAL MASS IS AT THE AXLE OF THE MAIN ROTOR,
THE VELOCITY OF THE MASS IS SUBTRACTED FROM THE VELOCITY OF THE RIM.
THEREFORE, CENTRIFUGAL ACCELERATION PREDOMINATES ON THE ONE SIDE.

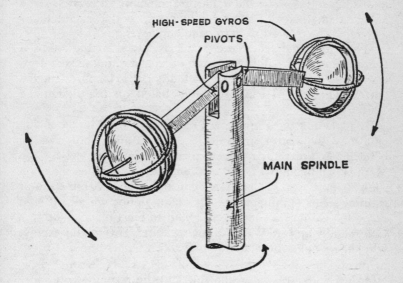

HIGH-SPEED GYROS

PIVOTS

MAIN SPINDLE

ESSENTIAL DESIGN
OF THE LAITHWAITE ENGINE

WHEN THE MAIN SPINDLE ROTATES,
PRECESSIONAL ACCELERATION
CAUSES THE GYROSCOPES TO RISE
AND FALL DURING EACH REVOLUTION.
THE MECHANICAL PROBLEM IS TO
ENGAGE THE RISE OF THE
GYROSCOPES TO GENERATE A LIFT
WHILE DISENGAGING THE
DOWNWARD SWING.

an implementation of the fourth principle of locomotion.

What Laithwaite needed was another rotary component, like the Dean Drive, geared to his engine's oscillations so that they would always be turned to drive in the same direction. As it happens, an Italian by the name of Todeschini recently secured a patent on this idea, and his working model is said to be attracting the interest of European engineers.

When the final rectifying device is added to the essential Laithwaite design, all the moving parts generate the vectors of a vortex, and the velocity generated is the axial thrust of the vortex. Therefore I call inventions based on this design the Vortex Drive.

By replacing the Hyperspace modules of the Mark I Flying Saucer with Vortex modules, still retaining the essential betatron as the centrifuge, performance is improved for the Mark II. To begin with, drive is generated only when the main rotor is revolving, so the saucer can be parked with the motor running. This eliminates the agonizing doubt we all suffered when the Lunar Landers were about to blast off to rejoin the Command Capsule: Will the engine start? This would explain why the ring of lights around the rim of a saucer is said to begin to revolve immediately prior to lift-off. A precessional drive affords a wider range of control, and the responses are more stable than a direct centrifuge. But the most interesting improvement is the result of the *structure* of the electromagnetic field generated by the Vortex Drive. By amplifying and diminishing certain vectors harmonically, the Mark III Flying Saucer can ride the electromagnetic currents of the Earth's electromagnetic field, like the jet stream. And this is just what we see UFOs doing, don't we, as they are reported running their regular flight corridors during the biennial tourist season. Professor Laithwaite got all this together when he conceived of his antigravity engine as a practical application of his theory of 'rivers of energy coursing through space'; he just could not get it off the drawing board the first time.

The flying saucer consumes fuel at a rate that cannot be supplied by all the wells in Arabia. Therefore we have to assume that UFO engineers must have developed a practical and compact atomic fusion reactor. But once the Mark III is perfected, another fuel supply becomes attainable, and no other is so practical for flying saucers. The Moray Valve will draw all the

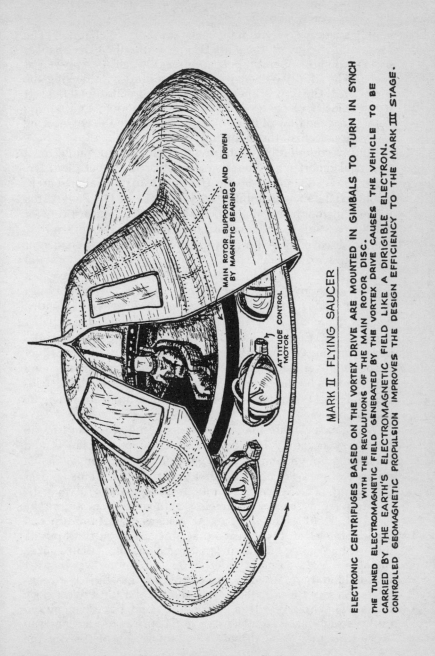

MARK II FLYING SAUCER

MAIN ROTOR SUPPORTED AND DRIVEN
BY MAGNETIC BEARINGS

ATTITUDE CONTROL
MOTOR

ELECTRONIC CENTRIFUGES BASED ON THE VORTEX DRIVE ARE MOUNTED IN GIMBALS TO TURN IN SYNCH
WITH THE REVOLUTIONS OF THE MAIN ROTOR DISC.

THE TUNED ELECTROMAGNETIC FIELD GENERATED BY THE VORTEX DRIVE CAUSES THE VEHICLE TO BE
CARRIED BY THE EARTH'S ELECTROMAGNETIC FIELD LIKE A DIRIGIBLE ELECTRON.
CONTROLLED GEOMAGNETIC PROPULSION IMPROVES THE DESIGN EFFICIENCY TO THE MARK III STAGE.

energy a saucer needs from the space through which it flies. The Moray Valve converts the Mark III into a Mark IV Flying Saucer by extending its operating capabilities through *time* as well as space. The Moray Valve, you see, functions by changing the direction of the flow of energy in the sun's gravitational field. It is the velocity of energy that determines motion, and motion determines the flow of time. We shall continue the engineering of flying saucers in the following essays.

My investigation into antigravity engineering brought me a technical report while this typescript was in preparation. Dr. Mason Rose, President of the University for Social Research, published a paper describing the discoveries of Dr. Paul Alfred Biefeld, astronomer and physicist at the California Institute for Advanced Studies, and his assistant, Townsend Brown. In 1923 Biefeld discovered that a heavily charged electrical condensor moved toward its positive pole when suspended in a gravitational field. He assigned Brown to study the effect as a research project. A series of experiments showed Brown that the most efficient shape for a field-propelled condensor was a disc with a central dome. In 1926 Townsend Brown published his paper describing all the construction features and flight characteristics of a flying saucer, conforming to the testimony of the first flight witnessed over Mount Rainier twenty-one years later and corroborated by thousands of witnesses ever since. (The Biefeld-Brown Effect explains why a Mark III rides the electromagnetic jet stream.)

We may speculate that flying saucers spotted from time to time may not only include visitors from other planets and travelers through time, but also fledglings from an unknown number of cuckoos' nests in secret experimental plants all over the world. The space program at Cape Canaveral may be nothing more than a supercolossal theater orchestrated by Cecil B. DeMille to reassure Americans that they are still *número uno* after Russia beat our atomic ace by putting Sputnik into orbit. We need not doubt that the Apollo spaceships got to the Moon, but we may wonder if Neil Armstrong was the first man to land there. The real space program may have been conducted in secret as a spin-off from the Manhattan Project since the end of World War II, and Apollo 13 may have been picked up by a sag wagon to make sure our team scored a home run every time they went to bat. The exploration of space is the most dangerous enterprise ever taken on by a living species. Don't you ever wonder why the Russians are losing men in space like a safari being decimated in headhunter country, while nothing ever happens to our boys except accidents during ground training?

5 The Philosophers' Stone

How to Transmute the Elements by Engineering the Geometry of Standing Waves

A series of experiments has been carried out in Japan proving that chickens fed a diet deficient in calcium produced, as the end product of their biological processes, *more* calcium than they were given to live on. The conclusion is that the chickens *created* the calcium they needed by transmuting potassium.

This discovery challenges the basic concepts of science, and the more critically a discovery challenges the foundations of scientific belief, the less it is examined at all. But if potassium can be transmuted into calcium (and by chickens, no less!), we had best construct a new model of the atom to explain how this might be possible. So let's get started, at the level of the subatomic particles that seem to be giving theorists so much difficulty.

After observing that light travels in straight lines to cast sharp shadows, Isaac Newton deduced that light beams could exist only if radiant energy possessed the characteristics of atomic particles. But Sir Isaac went on to pass beams of light through prisms, observing the spectrum of colors projected. The fractioning of light into colors is possible *only* if radiant energy possesses the properties of waves. The problem became a matter of determining whether light was particulate or wavy in nature. Theorists decided that the ultimate elemental substance was both particle *and* wave, depending upon what it happened to be doing with itself when observed. Then realists proceeded to advance science without caring what light was. Nevertheless, the problem for the philosophers remains. The properties of particles categorically exclude the properties of waves, so how is it possible for an elemental substance, whatever it is, to manifest both properties in succession?

After the greatest scientists since Newton have given up, all a lay person has to do is take a couple of cartons of quarter-inch ball bearings to a billiard parlor, rent a table, and spread balls on the baize.

84

After you have managed to arrange them with a mathematically random distribution, you will see that each ball is equally distant from its neighbours. Absolute chaos is identical to perfect order.

Now try to rearrange the balls so that groups are allowed, but the groupings are mathematically random. Eventually the pattern formed by the balls will follow a density of distribution described by the Bell Frequency Curve of random statistics. The Bell Frequency Curve is a sine-wave form; on a plane surface it is manifest as regular clusters, with small groups of roughly equal numbers being roughly equal distances apart. The smaller groups congregate into larger groups until the entire field can be described as a single sine-wave form of low frequency. Once again, you prove that utter disorder is identical to total organization.

If the balls are small enough and numerous enough in relation to the area you have to spread them on, you will discover the aggregations of particles will assume the pattern of a spiral generated by phi, the ratio between successive numbers in a series extended by adding consecutive numbers together; it is the ratio of 1:1.1618. All natural growth eventually follows the form of a spiral generated by a phi ratio, from the distributions of atoms to the distributions of stars in galaxies. (In other words, the spiral structure of gas clouds in interstellar space is not necessarily due to the process of gravitational contraction and centrifugal force, as proponents of the Nebular Hypothesis of stellar generation would have us believe. The spiral structure is an inevitable consequence of random distribution.)

You can perform this experiment at less cost by making pencil dots on a large piece of paper, but you will be bothered by constant erasing until you get the dots distributed properly. With pencil and paper, however, you can perform the converse experiment. Draw lines at random, each line representing a wave front. If you have enough lines on enough paper, and enough randomness, the result will look exactly like the random distribution of balls on a billiard table, as the intersections of lines form groupings of density.

Whether you perceive a ball to be an atomic particle or an aggregation of particles depends upon the *scale* of your frame of view. Whether you perceive an aggregation to be a particle or a wave depends upon the *scale* of your resolution. At the limit of resolution, all structures register on all instruments of measurement as particles. And all structures that cannot be resolved sharply by the instrument of measurement register as waves. So the nature of the ultimate element is determined by the instruments of measurement; all we can really know about it is what our instruments measure. Whether you choose to interpret reality as waves or particles depends entirely on what you want to do. The manifestations of energy—i.e., motion—yield measurements as waves; the manifestations of static material yield measurements as particles.

As it happens, everything is moving. Therefore, all events yield accurate measurements only as wave functions. The use of the laser for measurement establishes the wave as the elemental unit of space, time, motion, and energy.

As when Pythagoras studied music, harmonics is still taught from the model of a vibrating string. A plucked string vibrates back and forth as a unit, forming a standing-wave structure, emitting vibrations through the air to be heard as a musical sound. The tone is the fundamental frequency of the standing wave.

As the string vibrates as a unit, it also divides itself into two halves along its length, and each half vibrates as two individual standing waves independently of the fundamental wave. The frequency of the half lengths is twice the frequency of the fundamental, and the sound emitted is the second harmonic overtone, an octave higher than the fundamental.

And at the same time as the string vibrates as a unit and as independent halves, it also divides its length into three equal parts, each third vibrating independently to emit a sound three times the frequency of the fundamental, called the third harmonic overtone. At the same time, the string also divides itself into fractional lengths of quarters, fifths, sixths, and so on to the elemental molecular unit of vibration, generating successively higher harmonic overtones all the way. The distribution of energy among the overtones determines the unique sound characteristic of each instrument. This is the way harmonics is taught.

Only one thing is wrong with the course of study: The instructors got it all backward, just as electricians are taught that electricity flows in the opposite direction from the way it really flows. Now all the musicians and acoustic engineers will protest; everyone can see the vibrating string, and the course in harmonics describes exactly what you see—doesn't it? No, what

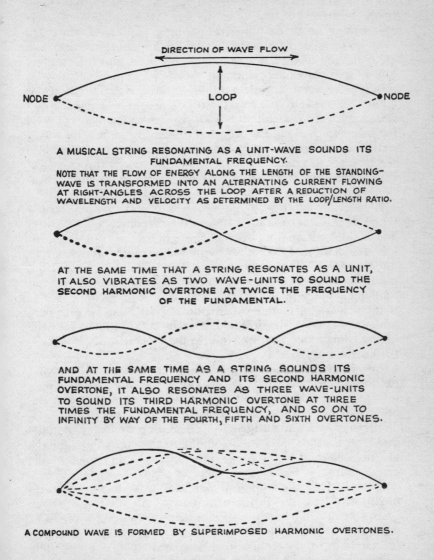

DIRECTION OF WAVE FLOW

NODE

LOOP

NODE

A MUSICAL STRING RESONATING AS A UNIT-WAVE SOUNDS ITS
FUNDAMENTAL FREQUENCY.
NOTE THAT THE FLOW OF ENERGY ALONG THE LENGTH OF THE STANDING-
WAVE IS TRANSFORMED INTO AN ALTERNATING CURRENT FLOWING
AT RIGHT-ANGLES ACROSS THE LOOP AFTER A REDUCTION OF
WAVELENGTH AND VELOCITY AS DETERMINED BY THE LOOP/LENGTH RATIO.

AT THE SAME TIME THAT A STRING RESONATES AS A UNIT,
IT ALSO VIBRATES AS TWO WAVE-UNITS TO SOUND THE
SECOND HARMONIC OVERTONE AT TWICE THE FREQUENCY
OF THE FUNDAMENTAL.

AND AT THE SAME TIME AS A STRING SOUNDS ITS
FUNDAMENTAL FREQUENCY AND ITS SECOND HARMONIC
OVERTONE, IT ALSO RESONATES AS THREE WAVE-UNITS
TO SOUND ITS THIRD HARMONIC OVERTONE AT THREE
TIMES THE FUNDAMENTAL FREQUENCY, AND SO ON TO
INFINITY BY WAY OF THE FOURTH, FIFTH AND SIXTH OVERTONES.

A COMPOUND WAVE IS FORMED BY SUPERIMPOSED HARMONIC OVERTONES.

is really happening is random motion. Whether or not you can hear the vibrations of a musical string above the audible threshold, the string is *always* vibrating due to the random molecular agitation of heat. (As far as the string is concerned, the extra vibration it gets from being plucked is just more heat.) Molecular motion along the string arranges itself into increasingly longer sine waves according to the Bell Frequency Curve of random distribution, until all the various fractional vibrations come into phase to generate the fundamental frequency. Fractions which do not coincide with the lower harmonics travel back and forth along the length of the string as moving waves until they come into direct opposition, transforming them into electromagnetic radiation. It's the loss of energy through electromagnetic transformation that causes the molecular vibrations to die down.

Electricians continue to learn their subject backward because which way the current flows makes no difference to the wiring; and besides, alternating current flows both ways. So what difference does it make whether harmonics is taught as division or integration? Well, as long as you believe electricity flows from positive to negative, you will never be able to discover and implement electronics.

If you learn harmonics by distributing ball bearings on a pool table, the way Pythagoras did after he was initiated into the higher dimensions and forswore beans, you will discover how the universe unfolds.

An infinite number of particles distributed and moving randomly through infinite space will divide themselves along a fundamental axis; one half moving in one direction and the other half moving in the opposite direction. This flow corresponds to gravity and antigravity. The reason we rarely see antigravity is that all particles belonging to the opposite pole have already departed in the other direction, and very few are left around here.

Each half of the universal particles traveling in opposite directions along the fundamental axis will divide into two groups again, moving in opposite directions along a plane at right angles to the gravitational axis. This secondary harmonic corresponds to the centrifugal and centripetal forces. The second harmonic will also subdivide into another pair of equal and opposite accelerations that can be represented as a cylinder parallel to the centrifugal-centripetal plane. The tertiary harmonic corresponds to the precessional forces.

Like the conventional view of the musical string, the universe can be described as subdividing itself successively until the ultimate particle is reached—whatever that ultimate particle is.

Of course, the universe does not really divide itself in this manner any more than the musical string does. It assembles its harmonics from random motion to coherent undertones. We proceed to analyze from the fundamental to the overtones only because it is convenient for our habit of thinking. We shall never know where the universal fundamental axis is, nor what the ultimate particle is, because in an infinite universe we must always find ourselves exactly in the middle of an infinite extension in both directions of whatever dimension we happen to be considering. What we call gravity, centrifugal-centripetal, and precessional forces are merely arbitrary conventions established for the convenience of our habitual mode of perception.

Once we perceive that all parts of space contain an indefinite number of particles moving at random to form the force fields we are familiar with, we understand how to engineer field energies directly. You see, one phase of precessional acceleration proceeds in the same direction as antigravity. To invent an antigravity engine, therefore, all you have to do is amplify the centrifugal harmonic until the antigravitational phase of the precessional harmonic exceeds the acceleration of gravity, and then eliminate the gravitational phase. This is exactly what Professor Eric Laithwaite calculated; he failed only because of errors in arithmetic. Other engineers have found the errors and corrected them. Whether or not the Laithwaite Engine worked, the fact remains that all antigravity engineering and all other field engineering can be reduced to the geometry of harmonics generated by random particles.

An infinite universe defined by an infinite number of randomly moving particles establishes the scientific principle of parity, meaning that energy will be equal in all directions and at all locations. In current physics, the concept of the cosmic hologram is still not accepted, so parity is limited to equality of motion in all directions.

When all the vectors of the gravitational-fundamental vibration and the centrifugal-centripetal secondary harmonic and the precessional tertiary harmonic and all the other harmonics are integrated into a resultant, the trajectory of any given particle must follow the course of a spiral vortex with a phi generator.

Therefore, any part of space you choose as a frame of reference will be defined by a fundamental field vortex, subdivided into an indefinite number of harmonic overtone vortices.

A vortex can spin in only one direction. Parity demands that for every vortex there must be a countervortex. This is why all dynamic structures are created, like men and women, in equal and opposite numbers.

Two vortices spinning in the same direction flow in opposite directions along their interface. Therefore, if pushed together, they will annihilate each other. This is why when particle meets antiparticle, they are transformed into radiant energy. The

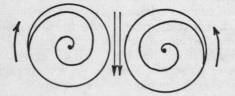

A RANDOM DISTRIBUTION OF PARTICLES IN UNIVERSAL SPACE CREATES VORTECES IN PAIRS TURNING IN OPPOSITE DIRECTIONS.
AT THE INTERFACE BETWEEN CONTRAROTATING VORTECES, THE FLOW IS IN THE SAME DIRECTION, SO OPPOSITELY TURNING VORTECES TEND TO MERGE.

AT THE INTERFACE BETWEEN VORTECES TURNING IN THE SAME DIRECTION, THE FLOW IS IN OPPOSITE DIRECTIONS, SO VORTECES TURNING IN THE SAME DIRECTION TEND TO PUSH AWAY FROM EACH OTHER.

THE BLACK HOLES THROUGH WHICH THE VORTEX FLOWS INTO HYPERSPACE IS THE OPPOSITE POLE OF THE CIRCLE FROM WHICH THE VORTEX FLOWS INTO THIS SPACE. THE BLACK HOLE AND THE WHITE CIRCLE ARE OPPOSITE SIDES OF THE SAME GATE TO AND FROM HYPERSPACE. THE EXPANSION OF THE POINT TO THE CIRCLE WITH SPACE SEPARATING THEM IS THE EFFECT OF PROJECTION BETWEEN HYPERSPACE AND REAL SPACE.

closer two vortices spinning in the same direction are pushed together, the more energy is brought into opposition along their interface. Therefore, all vortices rotating in the same direction will tend to move away from each other until they are spaced equally apart.

Conversely, two vortices spinning in opposite directions are flowing in the same direction along their interface. Therefore, they tend to merge. But they are not drawn together so much as pushed together by the pressure of similar vortices.

It is evident that the mechanics of vortices determines the force physicists call charge. Spin determines polarity.

Physical experiment has proven conclusively that electrons and protons are monopoles. The fact that electric charge is monopolar while magnetic charge is dipolar is one of the problems in the search for a Unified-Field Theory. If the vortex model is valid, however, electrons and protons should be dipolar, depending on which way they are oriented. But protons *always* repel each other, so all respectable physicists are convinced the vortex model is mistaken.

But protons do *not* always repel each other! When they come close enough together, they cleave together with greater force than any glue known. Physicists call this attraction the nuclear force, and they are unable to explain why it can be so powerful, but only over extremely close distances, within the nucleus of the atom. The answer is self-evident by a simple experiment. If you float a number of bar magnets in a fluid medium, and enclose the experimental setup in an electromagnetic field, the field will align all the magnets in the same direction and they will repel each other—like protons. But if the magnets are small enough, and are brought closely enough together, the mutual attraction of their opposite poles will overcome the force of the external field keeping them aligned—and they will flip, one relative to the other. With opposite poles tightly together, they will cleave together most tenaciously over a short distance. But once separated beyond the critical distance, the external field will align them in the same direction, and they will repel each other again. Scientists have come to perceive the electromagnetic field aligning particles in an atom as *the* electromagnetic field, so when particles flip and join in the nucleus with a thousand times more force than the attraction between proton and electron, a radically new force is postulated.

As it happens, Immanuel Velikovsky proposed an equivalent

hypothesis to explain why planets in the Solar System do not collide. You see, if there is mutual gravitational attraction among the planets, they must clump together over the course of time. But observations prove that the planets maintain the greatest possible distance from each other. When an extreme condition is maintained indefinitely, you cannot explain it as accidental; there must be a physical force keeping the planets apart. Unfortunately, it was Velikovsky who proposed this hypothesis, and no scientist who is not independently wealthy and careless of reputation can afford to prove anything that Velikovsky said.

Field forces are defined, by many criteria, so physicists may be on firm ground when they establish a nuclear force distinct from the electromagnetic force—but the experiments proving electrical particles to be monopolar do not contribute to that support.

When harmonic calculations are transferred to spaces of more than one dimension (the musical string is the standard object lesson), the same principles are assumed to be valid. As a consequence, spherical harmonics is interpreted as a circular wave expanding from a point of origin on the global surface, and the harmonic ratios are measured along a radius. This conception works very well as far as it goes, but as you will learn, plane harmonics has some extremely practical differences from linear harmonics.

A plane cannot exist as a vibrating structure unless it has at least three sides. The triangle, therefore, must be established as the fundamental unit of plane harmonics. When the sides of an equilateral triangle are bisected and joined, the result is four triangles, just as a square makes four squares when its sides are bisected and joined. The operations of plane harmonics apparently observe the rules of plane geometry.

William H. Whamond, writing in *Pursuit*, pointed out that if the sides of a polygon are not of a ratio that mutually reinforces each other's vibrations, the plane structure will disintegrate. All equal-sided polygons maintain their sides by mutual reinforcement, but all those which cannot be triangulated in harmonic ratios must collapse under pressure. It is surprising that Buckminster Fuller was able to build a career without realizing the function of harmonics in maintaining basic stability of structure.

Whamond went further to point out that although stabilizing the dimensions of diagonals may be sufficient for practical structures, theoretical requirements are not satisfied unless the diametric vibrations reinforce the perimetric vibrations to establish the polygon's rigidity through and through. The simplest polygon generated by a mutually harmonic reinforcement of both sides and center is the hexagon. This is the probable reason why six acquired a reputation for being the perfect number among the ancient philsophers, and why a circle's circumference was accepted as being three times its diameter, although every wheelwright knew better.

If you draw a grid of squares, and then draw all the diagonals, you will find yourself with a grid composed of two sets of squares. One set is rotated forty-five degrees from the other, and their dimensions are related to each other by a ratio of $\sqrt{2}$. This self-evident transformation assumes engineering significance when harmonic structures extend into higher dimensions.

As Buckminster Fuller pointed out, not only is the triangle the basic unit of plane space, but the principle of triangulation also establishes the tetrahedron as the basic unit of solid space. Like the triangle, however, the tetrahedron maintains its structure only by the triangulated rigidity of the mutually reinforcing vibrations of its sides. In order to possess internal stability, the tetrahedron must be doubled, one intersecting another, with points aligned on a polar axis.

To establish stability, tetrahedrons must always be manifest in mutually opposed and supporting pairs in this way. When this geometrical structure takes form from universal vibrations, however, it is not the simple pair of tetrahedrons it appears to be at first sight.

If lines are drawn joining all the points of the paired tetrahedrons, you have a cube. If lines are drawn between the centers of each face of the cube, they form the edges of an octahedron.

If circles scribed around the bases of the two tetrahedrons are divided into five equal arcs and all the points joined by lines, a symmetrical polyhedron defined by twenty equilateral triangles is defined. If all the points are joined by lines through the center of the icosahedron, twenty equal tetrahedrons will be defined. The space defined by an icosahedron is stabilized by mutually reinforcing resonance around all sides, along all edges, and through all diameters. Like the hexagon, the twenty faceted icosahedron is the perfect solid.

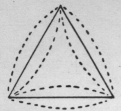

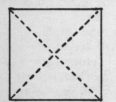

A TRIANGLE IS THE MINIMUM UNIT OF PLANE SPACE, DEFINED BY THREE WAVES IN A MUTUALLY REINFORCING STRUCTURE.

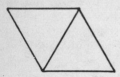

THE SQUARE IS CREATED BY FOUR MUTUALLY REINFORCING VIBRATIONS, BUT THE DIAGONALS ARE OF A RATIO INCOMMENSURABLE WITH THE SIDES IN WAVE UNITS.

THEREFORE, THE SQUARE TENDS TO COLLAPSE INTO A PAIR OF MUTUALLY REINFORCING TRIANGLES.

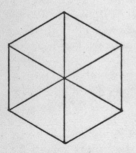

THE HEXAGON IS THE SIMPLEST, SYMMETRICAL PLANE SPACE ESTABLISHED BY A MUTUAL REINFORCEMENT OF THE DIAMETRIC WAVES WITH THE PERIMETRIC WAVES.

Now, bisecting all the lines forming an icosahedron produces a twelve-sided symmetrical polyhedron called a dodecahedron—the solid projection of the five-pointed star circumscribed by a pentagon. All ratios of the dodecahedron approximate the values of various mystical triangles, but they are incommensurate with the icosahedron by integers; the internal structure of the dodecahedron is irrational, like pi, phi, $\sqrt{2}$, $\sqrt{3}$,

94

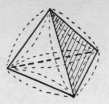

THE TETRAHEDRON IS THE MINIMUM UNIT OF VOLUME, DEFINED BY SIX STANDING-WAVES UNITED INTO A STRUCTURE OF MUTUALLY REINFORCING VIBRATIONS.

THE PRINCIPLE OF PARITY AROUND AXES ALIGNED AT RIGHT-ANGLES REQUIRES TETRAHEDRONS TO BE PAIRED.
A PAIR OF TETRAHEDRONS SHARING THE SAME BASE CONSTITUTES THE REGULAR TETRAHEDRON.

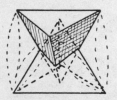

A PAIR OF TETRAHEDRONS INTERPENETRATING EACH OTHER POINT TO BASE IS STABILIZED BY A SET OF SIX WAVE UNITS RESONATING BETWEEN THE SIX EXPOSED VERTECES TO FORM A VIRTUAL OCTAHEDRON.
INTERNAL RESONANCE INTEGRATES WITH MUTUALLY REINFORCING EXTERNAL DIMENSIONS TO CREATE A POLYMORPHOUS HARMONIC STRUCTURE.

$\sqrt{5}$, etc. As you know, the diagonal of a square is related to its sides by $\sqrt{2}$, and the diagonal cross of a square is also the *negative* of the square. The dodecahedron is the negative of the icosahedron. In this context, the octahedron is the negative of the cube. Because a tetrahedron is the elemental unit of solid space, no other polyhedron can function as its inferior negative, so the tetrahedron is rotated 180 degrees to function as

THE PRINCIPLES OF PARITY, HARMONIC
INTEGRITY AND ROTATION TOWARDS THE
SIMPLEST POSSIBLE ISOMETRIC STRUCTURES
UNITES TETRAHEDRONS IN POLAR OPPOSITION
BETWEEN FOUR PAIRS OF VERTECES,
INTERPENETRATING EACH OTHER AT MIDSECTION.
EACH EDGE IS DIVIDED INTO HALVES.
EACH POLAR AXIS IS DIVIDED INTO THIRDS.
A POLAR PROJECTION SHOWS THE
EQUATORIAL PLANE DIVIDED INTO SIXTHS.
EACH POINT IS A TETRAHEDRON DEFINED
BY EDGES ONE-HALF THE LENGTH OF
MAIN TETRAHEDRONS.

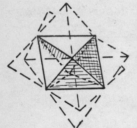

WHEN THE TETRAHEDRONS FORMING
THE VERTECES OF THE EIGHT-POINTED
SOLID STAR ARE CUT OFF, THE CORE
REMAINING CONSTITUTES AN
OCTAHEDRON.
THE HARMONIC INTEGRATION OF
QUARTERS AND EIGHTHS WITH
HALVES AND THIRDS IS REVEALED

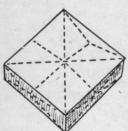

WHEN THE VERTICES OF THE PRIME
TETRAHEDRONS ARE JOINED BY LINES,
A CUBE IS DEFINED.
THE VERTECES OF THE OCTAHEDRON
DEFINE THE CENTRES OF THE SQUARES
FORMING THE CUBE AND EACH EDGE
OF THE PRIME TETRAHEDRONS ESTABLISHES
THE DIAGONALS OF THE SQUARES.
NOTE THAT THE AXES OF EACH SOLID
HARMONIC IS ROTATED RELATIVE TO
THE STRUCTURE WHICH GENERATES IT.
THIS PROVES THAT A GIVEN SET OF
VIBRATIONS DEFINES THE FIELD IN SEVERAL
DIFFERENT STRUCTURES AT THE SAME
TIME IN THE SAME PLACE AS DETERMINED
BY THE PHASE HARMONICS INTEGRATING
THE SET. A GIVEN STRUCTURE BECOMES
TANGIBLE WHEN ITS PHASE IS
TUNED TO THE PERCEIVER.

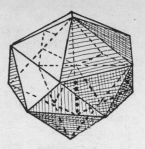

WHEN CIRCLES DRAWN AROUND THE BASES OF A PAIR OF MUTUALLY OPPOSED TETRAHEDRONS ARE DIVIDED INTO FIVE EQUAL SEGMENTS AND THE POINTS JOINED BY LINES, TWENTY EQUILATERAL TRIANGLES ARE FORMED AS THE FACETS OF A REGULAR ICOSAHEDRON. THE TRIANGLES ARE THE BASES OF TWENTY REGULAR TETRAHEDRONS RADIATING FROM THEIR APECES AT THE CENTRE OF THE ICOSAHEDRON.
THE ICOSAHEDRON IS THE SOLID CORRELATIVE OF THE PLANE HEXAGON, BEING STABILIZED BY DIAMETRIC VIBRATIONS MUTUALLY REINFORCING ITS SURFACE VIBRATIONS.

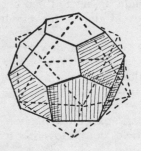

LINES CONNECTING THE CENTRES OF THE FACETS OF THE ICOSAHEDRON DEFINE THE EDGES OF TWELVE PENTAGONS FORMING THE FACETS OF A REGULAR DODECAHEDRON. THE BISECTING LINES ARE THE CENTRAL AXES OF THE TRIANGLES, RELATED TO THE SIDES AS $\sqrt{3}$ IS TO 2. THEREFORE, THE CENTRAL AXES OF THE TRIANGLES ARE THE NEGATIVES OF THE TRIANGLES AND THE DODECAHEDRON IS THE NEGATIVE OF THE ICOSAHEDRON. CORRELATIVELY, THE OCTAHEDRON IS A PARTIAL NEGATIVE OF THE CUBE.

its own solid negative. Lines drawn from the points of a tetrahedron to its internal center form a linear structure called the Miraldi angle, resembling a caltrop; this is the true negative of the tetrahedron, but the field rotation required to transform a tetrahedron into a caltrop projects the structure into fewer dimensions.

Now, the relationship between a square and its diagonals is a 45-degree rotation on plane space, which is the projection of a rotation of 90 degrees in hyperspace. The relationship between the tetrahedron, the octahedron, the cube, the dodecahedron, and the icosahedron is also established by a definite rotation through hyperspace. The notes of a musical scale are also defined by a definite rotation of energy through hyperspace, which transforms one frequency into another. As an illustration, rotation through hyperspace transforms the wavelength of the side of a square to the wavelength of its diagonal. The ratio of the side to the diagonal is the same as the ratio between G and C on the musical scale.

You have just made a discovery sought by philosophers throughout history; the regular Platonic solids are related to each other as musical notes on a hexatonic scale. Extend the sides of the dodecahedron until they meet, and you have the frame of a pair of tetrahedrons exactly twice the size of the pair you started with to continue the scale on the second octave. You have discovered the Music of the Spheres.

Parity is not satisfied by the creation of nuclear particles in the form of equal and opposing standing-wave vortices. The axes of the pair, you see, are both aligned in the same direction; that is a manifestation of directional preference. In order that axes be balanced in all directions, particles must congregate in groups of six: three pairs of mutually opposed particles with the axis of each pair at right angles to the other two. This assemblage puts each vortex at the vertex of an octahedron.

The octahedron is not stable because each pair of vortices grind gears against the other two. But if the equatorial pair of particles move away from each other along the polar axis, the six can mesh together like two pairs of crown gear clusters fitted at right angles to each other.

The vortex model suggests that the basic particle is likely to be composed of three pairs of finer particles bound together in

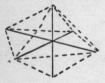

THREE AXES ALIGNED AT RIGHT-ANGLES TO EACH OTHER ESTABLISH THE SIMPLEST STRUCTURE POSSIBLE IN 3-D SPACE. THE NEGATIVE (HYPERSPACIAL) AXES GENERATING THE OCTAHEDRON CONFORM TO THIS LIMITING CASE OF SIMPLICITY. THIS IS WHY RIGHT-ANGLE COORDINATES ARE FUNDAMENTAL TO GEOMETRY INSTEAD OF THE TRIANGULATION FAVOURED BY BUCKMINSTER FULLER.
THE OCTAHEDRON IS THE SIMPLEST STABLE ELEMENT OF GEOMETRY.
THE TETRAHEDRON IS THE THIRD HARMONIC OF THE BASIC OCTAHEDRON.

FOUR AXES ALIGNED AT 127° INTERVALS ARE NEEDED TO GENERATE THE TETRAHEDRON. THE PRINCIPLE OF PARITY GIVES LOW PROBABILITY TO TETRAHEDRONS EXISTING BY THEMSELVES; LIKE HYDROGEN ATOMS, THE PREFERRED STATE OF THE TETRAHEDRON IS IN PAIRS. THE SAME AXES WHICH GENERATE THE ELEMENTAL TETRAHEDRON ALSO GENERATE A PROPERLY OPPOSED PAIR WHEN EXTENDED TO DEFINE THE DIAGONALS OF A CUBE.

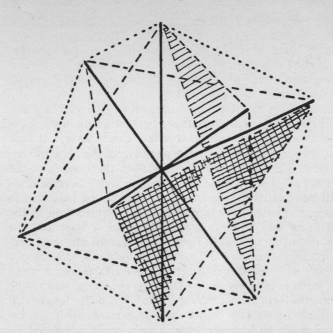

NOTE THAT THE AXES GENERATING THE OCTAHEDRON FORM THE CORE OF THE TETRAHEDRON PAIR.

THIS MODEL REVEALS THAT THE INTERSECTION OF HYPERSPACIAL FIELDS AT RIGHT ANGLES ON ALL AXES GENERATES OCTAHEDRONAL STRUCTURES IN REAL SPACE. WHEN FIELDS IN HYPERSPACE INTERSECT WITH AXES ALIGNED ALONG THE TERTIARY HARMONIC, CUBIC/TETRAHEDRONAL STRUCTURES ARE CREATED. BY ROTATING HYPERSPACIAL AXES, ALL CONCEIVABLE STRUCTURES CAN BE CREATED IN REALSPACE. THIS IS THE BASIS OF ALCHEMY, MAGIC AND WITCHCRAFT. THE MIND IS A HYPERSPACIAL ENTITY CAPABLE OF AUTONOMOUS ROTATION OF FIELD AXES. THE TRANSLATION OF BRAIN MECHANICS INTO TECHNOLOGY IS WHAT PSYCHOTRONIC ENGINEERING IS ALL ABOUT.

the harmonic structure of an octahedron. The geometry of the three pairs bears a striking correspondence to the characteristics of the elusive quark. Charm, beauty, and color appear to be manifestations of angle in hyperspace; axial angles account for fractional electric charge.

The octahedron still does not quite satisfy parity. The polar pinions of the crown gear clusters are both spinning in the same direction; this will give the octahedron a net charge. If four more pair of vortices, forming the negative of the octahedron, are spaced in a cube arrangement between the vertices of the octahedron, all the gears will spin in the right direction, all spins will be equally opposed, and all axes will be balanced in all directions. The cube-octa is the likely conformation of the neutron. Proof will be slow coming because at least half the particles are in the quantum field at the instant any measurement is made; this is why the quarks are so damned elusive.

The cube-octa contains fourteen particles. If struck, it could collapse, with twelve arranging themselves around one in the center, in the form of a dodecahedron, while the fourteenth spins free into orbit. The transformation is remarkably similar to what appears to happen when a neutron is converted by impact into a proton and an electron.

If the proton has the geometry of a dodecahedron, it will be a charged particle, so every proton will seek another proton as a mate. This may be why hydrogen is a diatomic molecule. After the neutron collapses, parity is not reestablished absolutely until the helium atom is formed. This would explain why helium is monatomic, with all the properties of an overgrown neutron.

Now that we have our electrons, protons, and neutrons straightened out, let's put them all together!

Niels Bohr described the atom as a miniature Solar System, with the nucleus serving as a Sun, orbited by electron 'planets'. The Bohr model is represented in all popular scientific literature despite the fact that any child can see it must be impossible. You see, if you have electrons orbiting in all directions around a nucleus, they are bound to collide; and on the atomic time scale, *eventually* is something sooner than a microsecond.

The atomic traffic problem was solved by giving each electron a different radius to orbit, but this solution won't work, either. An electron's wavelength is defined by its orbit. If every

100

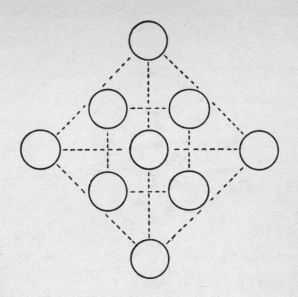

POLAR VIEW OF THE PROBABLE DISTRIBUTION OF ELECTRONS
IN THE OUTERMOST SHELL OF THE KRYPTON ATOM· THE
OPPOSITE HEMISPHERE IS THE SAME.
THE BASIC OCTAHEDRON COMBINED WITH THE CUBE
REESTABLISHES THE GEOMETRY OF THE NEUTRON ON A
LARGER SCALE.

PARITY OF THE MODEL BREAKS DOWN AT THIS STAGE OF
DEVELOPMENT. KRYPTON CAN BE MADE TO COMBINE CHEMICALLY
BUT ITS ACTIVITY IS FAR FROM COMMENSURATE WITH THE
FAILURE OF PARITY. THEREFORE, ANOTHER HARMONIC
FEATURE MUST BE INTRODUCED IF THIS MODEL IS TO
REMAIN VALID FOR HEAVIER ELEMENTS.

electron has its own orbital radius, each electron will manifest a different wavelength. This does not happen.

Erwin Schrödinger resolved the problem by proposing that electrons were standing waves, but his equations required three dimensions for each electron. Although the standing-wave equations were accepted, the necessity for multiplied space was not. As a consequence, mathematical physicists are still searching for a model that will make the atom possible! They have given up seeking a model that can be represented as a mechanical structure, and physics builds increasingly complicated and abstract equations.

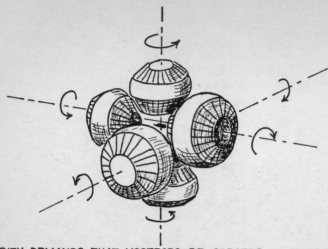

PARITY DEMANDS THAT VORTECES BE CREATED IN THREE
PAIRS OF OPPOSITELY SPINNING SPHERES WITH THEIR
AXES ALIGNED AT RIGHT ANGLES.
IN THIS DIAGRAM, THE POLAR VORTECES SPIN IN THE SAME
DIRECTION, VIOLATING PARITY.

The model of solid harmonics indicates that the node of the
electronic standing wave revolves around the equator of the
hydrogen nucleus. The node requires only half the quantum
orbital space it has, so another electron can share the same shell
to form a helium atom. To maintain parity, each moves to a
polar hemisphere separated by the equator, and revolves in
opposite directions.

Space is insufficient for a third electron, so the lithium atom
must start another shell. The second shell has enough area for
eight electrons, so the surface of each hemisphere is divided
harmonically into successive halves, thirds, and quarters.
Apparently the equatorial division establishes a hemispherical
sector that is never crossed. The eight facets form the sides of
an octahedron (the ubiquitous octahedron again) and each
facet has just enough room to hold an electron; each facet is a
quantum unit of space relative to the frequency of the electron.
When the octahedron is complete, the atom is electrically neu-
tral, as all octahedrons with their gears running smoothly are
supposed to be. Neon is almost as inert as helium, but parity
must be observed; and a second octahedron is laid over the first
with the spins of each electron aligned at a different angle. The

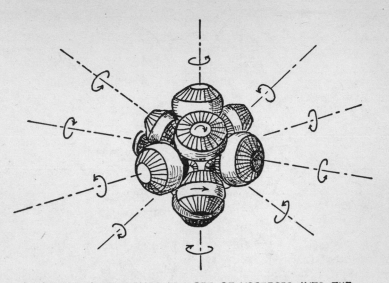

INTRODUCING AN INTERMEDIARY SET OF VORTECES INTO THE
NEGATIVE SPACE, FORMING A CUBE WITHIN THE OCTAHEDRON,
CORRECTS THE ROTATION SO THAT EACH POLAR VORTEX ON
EACH AXIS SPINS IN OPPOSITE DIRECTIONS. PARITY IS SATISFIED
BY ALL DIRECTIONS OF SPIN AND AXIAL ALIGNMENTS.

NOTE THAT THIS STRUCTURE IS A HARMONICALLY INTEGRATED
SOLID SCALE CORRESPONDING TO THE MUSIC OF THE SPHERES.

IF REALITY IS STRICTLY GOVERNED BY PARITY, AND IF
PARITY IS ESTABLISHED BY RANDOM VECTORS, THEN THIS
MODEL MUST BE THE PROBABLE STRUCTURE OF THE NEUTRON.

THE SPIN, MASS AND CHARGE OF EACH VORTEX DEPEND ON
ITS AXIAL ALIGNMENT TO THE FRAME OF REFERENCE AT THE
INSTANT MEASUREMENT IS MADE. THEREFORE, THE NATURE OF
SUBATOMIC PARTICLES MUST BE INDEFINITE IN THEIR VARIETIES.
THE ONLY CONSTANT IS THE FREQUENCY AND STABILITY OF
EACH HARMONIC STRUCTURE AS ESTABLISHED BY PROBABILITY.

fourth orbital shell has sufficient radius for its surface to hold
many more electrons. If you move the eight electrons to the
vertices of the octahedron harmonic structure, so that each
hemisphere is covered by a square pyramid with the equatorial
cleavage separating them, you will find enough room to add
another electron to the center of each facet, defining a cube-
octa. The total of electrons will be eighteen; this is the number
of electrons proven to be established in the subsequent shells in
the generation of the Periodical Table of Elements. The electrons of each shell align their axes to balance parity.

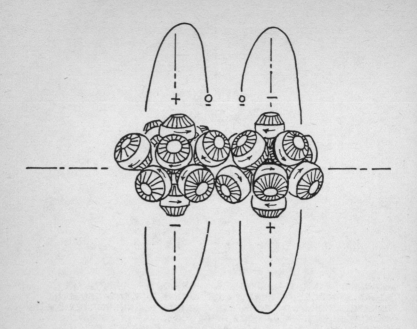

THE PROTON PROBABLY ASSUMES THE STRUCTURE OF THE
DODECAHEDRON. THE DODECAHEDRON ASSEMBLAGE OF VORTECES
HAS THE POLAR PAIR SPINNING IN THE SAME DIRECTION, GIVING
THE PROTON A NET CHARGE. THE CHARGE IS NEUTRALIZED
WHEN THE PROTON UNITS IN PAIRS ALIGNED IN OPPOSITE
DIRECTIONS CREATE THE HYDROGEN MOLECULE ILLUSTRATED
ABOVE. THE HYDROGEN MOLECULE STILL FAILS AXIAL PARITY,
SO HYDROGEN REMAINS CHEMICALLY ACTIVE. THE ADDITION
OF A PAIR OF NEUTRONS BETWEEN THE ELECTRONS ON AN
AXIS PERPENDICULAR TO THIS PLANE ESTABLISHES THE
AXIAL STRUCTURE OF THE OCTAHEDRON, MAKING THE
HELIUM ATOM UTTERLY INERT.

But apparently it is a long way from helium to the next perfect
atomic geometry. If the equatorial cleft is retained throughout
the generation of elements, the model of heavier atoms will
assume the dumbbell configuration of the electromagnetic field
surrounding a bar magnet. (Structural weakness at the waist
may be the reason that atoms heavier than bismuth break spon-
taneously.)

To illustrate how the geometry works in practice, the atom of
carbon has a pair of miter caps, one over each polar hemisphere.

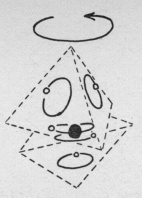

THE OXYGEN ATOM PROBABLY HAS THREE ELECTRONS
REVOLVING IN PLANES FORMING A TETRAHEDRON OVER
EACH POLAR HEMISPHERE. THE RADIUS OF REVOLUTION
IS THE SAME FOR ALL ELECTRONS IN ALL SHELLS SO
THAT THE FREQUENCY OF THE ELECTRON CAN BE IDENTIFIED
AS CONSTANT. THE SHELLS SPIN AROUND THE NUCLEUS,
SATISFYING THE EXPERIMENTAL BASIS OF THE BOHR MODEL.
THE STATE OF ATOMIC EXCITEMENT IS DETERMINED BY THE
VARIABLE RADIUS OF THE SHELL AS IT IS TRANSFORMED
FROM ONE HARMONIC OVERTONE TO ANOTHER.

There is space for two more electrons in each hemisphere to
complete the octahedron. When it takes the electrons attached
to hydrogen atoms, the hydrogen nucleus is going to stick out as
a lump. In order to maintain parity, the angles at which the
hydrogen atoms will join the carbon atom to form methane
conform to the points of a tetrahedron. This fact is taken for
granted in stereochemistry today, but established authorities
put down the first chemist who suggested that molecules had
solid structures, quite different from the empirical formulas
used to describe them.

The oxygen atom is capped by three-sided pyramids with
room for one more electron in each hemisphere. Parity allows a
120-degree angle between the hydrogen nuclei, and so water
forms ice crystals in a hexagonal geometry.

Outside of the innermost shells, electrons do not orbit the
nucleus of their atoms at all; they orbit the space of their
octahedronal facet at a constant radius. This geometry makes it
possible to avert collisions and maintain a constant frequency of

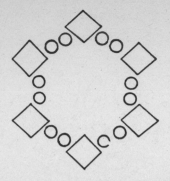

WHEN A HYDROGEN ATOM COMBINES WITH EACH POLAR
HEMISPHERE OF THE OXYGEN ATOM, THE HEXAHEDRON
GEOMETRY IS TRANSFORMED INTO THE STABLE OCTAHEDRON.
THE PLANES OF THE OCTAHEDRON MEET THE POLAR AXIS
AT AN ANGLE LITTLE MORE THAN 30°.
THEREFORE, WHEN WATER MOLECULES ARE CRYSTALLIZED
BY THE HYDROGEN BOND, THE SNOWFLAKE TAKES FORM
AS A HEXAGON.

orbit, regardless of an electron's distance from its atomic center. When atoms are excited by absorbing radiation, a rotation in hyperspace causes the shells to move out to a greater radius from the nucleus to the positions calculated from experiment.

You have been taken along this line of superficial physics and chemistry to give you a basis for the possibility that all molecular structures are generated from the elementary geometry of the Platonic solids, with the elements combined in various combinations of harmonically integrated angles, like crystals. If this is so, then each chemical element and compound will resonate in sympathy to a specific geometric solid. Furthermore, each solid structure can be excited and modulated by musical sound. This is not a novel concept, but the very basis of alchemy.

Now, each solid can be transformed into another structure by a regular rotation through the hyperspace of the quantum field. Each chemical atom is also transformed by a rotation of its geometrical structure in hyperspace. Therefore, by employing tuned vibrations it is theoretically possible to transform lead into gold (or gold into oil, which is considerably more valuable these days).

Fitting experimental data to the theory of solid harmonics is a task requiring professional competence. Even if the essential concept is correct, conflicting data is turning up day by day inspiring many false starts.

In the meantime, back at the bench, we have discovered the Philosophers' Stone. If a birdbrain can transmute the elements, so can engineering geniuses—as soon as we figure out how those stupid chickens did it.

6 Time Travel

*How to Navigate the Streams of Time
Through Hyperspace*

The technology dominating a civilization is determined by the concept of reality believed by its social leaders. Concepts of reality are axioms which cannot be tested. This is not to say that cultural beliefs are refractory to proof; it is impossible for people to examine their concepts of reality because there is a mental block at the depths of individual and collective psychology enforced by taboo. The concept of reality must be guarded inviolate from question. Merely to think about it begins to dissolve its absolute power over us and leads to the transformation of one person after another until the civilization disintegrates.

The fundamental and unquestionable concepts of our civilization are Absolute Time, Absolute Space extending in three dimensions, Absolute Velocity, and Conservation. These realities are so self-evident to us that we are unable to perceive evidence to the contrary. We cannot believe that people in other times and places do not perceive time and space as we do. If we could see the everyday life of the Middle Ages, as it was experienced by the average person who lived then, we should regard the entire population of medieval Europe as paranoid schizophrenic. When Einstein proved that time, space, velocity, and conservation were not absolute but mutable, the heresy was allowed to exist only because it was expressed in a language defying all comprehension.

Modern inquisitors state that the psychedelic experience provides no insight into the nature of reality, as the mystic heretics claim; the psychedelic experience is nothing but the hallucinations of a disordered mind. But Einstein pointed out that it takes only a single contradictory fact of evidence to demolish a scientific theory. So of all the various psychedelic experiences, we need only one to prove that the prevailing concepts of reality must be profoundly mistaken. It is fairly common for freaks to experience time flowing backward. Now, for time to be experienced in retrograde, it is necessary that a person have a clear and continuous memory of the future. If the concept of absolute time is true, then it is absolutely impossible for anyone to experience a memory of the future, no matter how disordered his mind may be. Therefore, time cannot be absolute.

Mystics are handicapped by lacking scientific concepts to explain their experience in scientific terms to scientific critics. Anyone who has passed high-school science, however, has all the scientific concepts needed to comprehend both the psychedelic experience and the Theory of Relativity, provided he is an acute observer and uses common sense instead of believing the professional gurus.

To show you how easy it is to break the taboo maintaining mistaken concepts of reality, Buckminster Fuller holds a global map of the world in his hand so that a triangle drawn on its surface can enclose 270 degrees of angle. The first man who got his observation of spherical geometry published in a hard-core scientific journal is regarded as a great mathematical genius, but everyone has always seen that gores on a curved surface enclose more or less than 180 degrees of angle. Tailors and sheet-metal mechanics have practiced spherical geometry since prehistoric times.

Spherical geometry leads directly to the Theory of Relativity and the transformation of our civilization by introducing radically different concepts of reality. The atomic bomb and atomic power are merely the first technological advances made possible by the new beliefs. A number of more or less simple models will enable you to understand the consequences of Relativity and the incredible technology already under study in laboratories for research and development.

Take a flat sheet of metal or vibrant plastic, sprinkle its top surface with a fine powder, and strum a bow across one of its edges. The sheet will resonate like a violin string. The sonic wavelengths which coincide harmonically with the dimensions of the particles of powder will cause the particles to vibrate sympathetically and bounce up and down. These vibrations carry the particles with them, making the pattern of the field of sound waves visible. This Plate Flutter Model illustrates the structure of the entire universe and shows how it unfolds.

Sound vibrations travel eleven hundred feet a second in air, faster in solids, so their movement is much too rapid for the eye to follow on a small plate. What you see are standing waves formed when radiant vibrations meet from opposite directions. It may help you to understand standing waves by studying waves on a body of water. Water molecules vibrate as waves traveling through the liquid with sonic velocity. Where these waves intersect to augment each other, a wave pattern is created that doesn't move. If enough sonic waves augment each other at the same location, the standing-wave pattern grows large enough to be seen as waves rising and falling on the surface of the water. A standing wave is like a stroboscopic image, whereby a high-speed, repeating pattern is made to appear to move slowly.

In an ideal experiment, the pattern of the standing waves would reveal only the structure of the sound vibrations. A real model, however, is limited to revealing only the patterns which conform to the resonance of the particles and the plate. Furthermore, sound waves are three-dimensional structures; they fill solid space. All the model can show is a plane section through the solid field of sound waves. These limitations must be borne in mind when reading the experiments.

You can change the frequency of the tuning by sounding another note. You can change the way the vibrations meet by cutting the plate to other shapes. If you use a chord for sounding, instead of a single note, a change of individual notes in the chord will change the patterns. If you use a tuning fork to apply sound to the plate, the patterns will change according to the position of the fork, and you can use more than one fork to alter tuning by varying their positions one to another. The powder will respond to all these changes by moving to form other patterns in other places.

Watch closely while tuning is changed. As opposing vibrations are shifted from perfect congruence, the standing-wave patterns move. As a standing-wave pattern acquires velocity, its extension in the line of travel contracts. What you are seeing is an illustration of the Fitzgerald Contraction described in Einstein's Special Theory of Relativity happening right in front of your eyes.

Time is defined by vibrations. If you inspect a standing-wave pattern intently, you will see that as it acquires velocity, it vibrates more slowly; the reciprocating motion is transformed into linear motion. What you are seeing is the slowing down of time as velocity is increased. In most cases, the motion will require high-speed movie film to make it fully visible.

A standing-wave structure manifests all the properties of material mass in physical equations. When a standing-wave pattern is accelerated to the velocity of the radiant waves which create it, its three-dimensional structure disappears altogether. What you are looking at now is the conversion of pure matter into pure energy, as described by Einstein in his famous equation, $E = mc^2$. An atomic bomb is nothing more than the explosive acceleration of solid atoms to the velocity of light, at which speed the material is transformed into radiant waves.

The model makes it obvious that no material structure can be accelerated to a speed greater than the speed of the waves which create it. This is why the speed of light is the limiting velocity for all material bodies.

If time slows down in your frame of reference as you acquire velocity, the time of the frame surrounding you appears to accelerate in comparison. The effect is the same as traveling through time into the future. The Theory of Relativity makes it theoretically possible to travel through time.

In fact, when you leave the city at the end of the day to spend the night in the suburbs, the city you return to is shifted into the future relative to where it would be if you spent the night in a downtown hotel. The shift is negligible, however, because the speed and duration of your commutation is nominal relative to the flow of time. If you could travel to the Sun and back at the speed of light, the Earth would progress sixteen minutes into the future while you were experiencing no passage of time at all. You would have to travel to the nearest star and back at the speed of light to step off your time shuttle ten years in the future.

Acceleration to the speed of light takes you on a one-way trip into the future. Without a return ticket, the trip is as attractive as dying. If acceleration transports you to the future, then you must travel at less than zero velocity to return to the past. It is possible to have less than nothing in your bank account by over-drawing, but moving slower than standing still seems impossible. Minus velocity only appears to be impossible, however, because of the way we think about speed. In the mathematics of relativity, less than zero velocity is equivalent to hyperlight velocity, like five to noon is merely another way of saying fifty-five minutes after eleven o'clock.

The logic of traveling backward in time is as simple as the logic of traveling forward into the future. You see, all acceleration must be through time as well as space. From zero velocity to light velocity, all travel must be into the future. There is only one way to go after the velocity of light is exceeded. That way is back into the past.

You have just seen the Plate Flutter Model demonstrate Einstein's proof that no material body can be accelerated above the speed of light, so how is it possible to attain the hyperlight velocities required to travel backward into time? Intensive analysis of the Plate Flutter Model will show you that although no standing wave can exceed the speed of light, the information

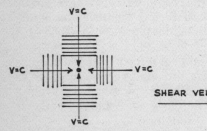

SHEAR VELOCITY

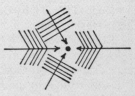

ALL MATERIAL BODIES ARE STANDING-WAVE STRUCTURES CREATED BY UNIVERSAL VIBRATIONS COMING TO A FOCUS FROM ALL DIRECTIONS IN PROPER PHASE AND RETURNING TO THE QUANTUM FIELD AT THE PARTICULATE CENTRE.

ALL BODIES MOVE AS THEIR PROPER WAVEFRONTS ROTATE ON THE PLANE OF THEIR FLOW. WAVEFRONT ROTATION IS DUE TO THE ROTATION OF THE UNIVERSAL FIELD AROUND THE AXES OF PHASE AND FREQUENCY.

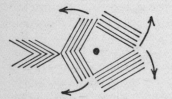

AS SHEAR VELOCITY APPROACHES THE SPEED OF LIGHT, THE TRAILING WAVEFRONTS LAG BEHIND THE MOVING BODY. AS THE VELOCITY OF THE SHEAR ANGLE OPENING THE LINE OF TRAVEL EQUALS AND THEN EXCEEDS THE FLOW VELOCITY OF THE WAVES, THE REDUCTION OF THE FREQUENCY OF THE WAVES CREATING THE ACCELERATING BODY IS MANIFEST AS A SLOWING OF TIME UNTIL TIME STOPS.
WHEN THE ROTATIONAL VELOCITY MOVING AWAY FROM THE FOCAL POINT EXCEEDS THE WAVE VELOCITY FLOWING TO THE BODY, THE WAVE VECTORS ARE REVERSED IN DIRECTION AND TIME FLOWS BACKWARD.

contained in the pattern of the standing wave can be accelerated to infinite velocity in the form of a shear-wave pattern.

A shear wave is illustrated by the angle formed by a pair of scissors. As the blades close, the point of the intersection accelerates to infinite velocity.

A material standing-wave structure is a highly compounded shear-wave structure. A standing wave is created when a pair of radiant waves meet at an angle of 180 degrees, interpenetrating each other. Three pairs of mutually opposed radiant waves must meet at the intersection of three axes, all at right angles to each other, to generate a solid-material body. The acceleration of a body in a field is the effect of a rotation of the flow axes of the radiant waves which generate it.

As the straight angle closes like a pair of scissors on one axis, the standing wave accelerates at right angles to the direction of radiant-wave flow. You can observe this in moiré patterns available from the Edmund Scientific Company, advertised in many technical magazines on the newsstand.

As a standing wave acquires shear velocity, it can be seen to disintegrate in the line of travel; its material dimensions are transformed into radiation according to the Fitzgerald Contraction. The radiation emitted ranges from the gravity waves broadcast by a massive body accelerated in a gravitation field to the synchrotron radiation beamed by an atomic particle accelerated in an electromagnetic field. This disintegration is apparent only to observers in the original frame of reference. To the accelerated body, it is the observers at the home base of reference that disintegrate into pure radiation.

As an accelerating body disintegrates along the axis of travel, it re-forms along an axis proceeding at right angles to the line of travel. As a result, the standing wave is observed to veer to the right as Relativistic velocities are reached. This is the reason for the right-hand rule learned by electrical engineers.

As an accelerating body acquires velocity on the second axis of travel, it begins to disintegrate in that direction also, and re-form along a third axis at right angles to the other two. As a consequence, all particles moving at Relativistic velocities are seen to travel in a corkscrew trajectory. The velocity of light is never quite reached in any direction because the trajectory curves over an angle of 45 degress every time the velocity on the relevant axis reaches $\frac{c}{\sqrt{2}}$.

The speed of light is reached on all axes simultaneously when all the flow axes of the waves which create the standing wave rotate through a total of 540 degrees. At this point, the 3 axes of generation close from 180 degrees to zero degrees simultaneously; the standing wave disappears over the event horizon of a Black Hole at the speed of light, and jumps to infinite velocity. At the speed of light, the traveling body is transformed into pure radiation. When the information contained in that radiation flips into the Black Hole, the waves themselves explode in all directions into the quantum field in the form of virtual particles.

If the coordinates of the moiré pattern are properly programmed into a computer to be projected onto a screen, a strange phenomenon will become manifest when the three generating axes become parallel. The standing-wave pattern is

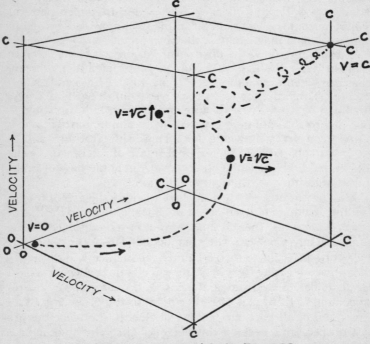

AS AN EVENT IS ACCELERATED, IT FOLLOWS THE RIGHT-HAND RULE GOVERNING ELECTROMAGNETIC VECTORS TO TRACE THE TRAJECTORY OF A SPIRAL VORTEX.

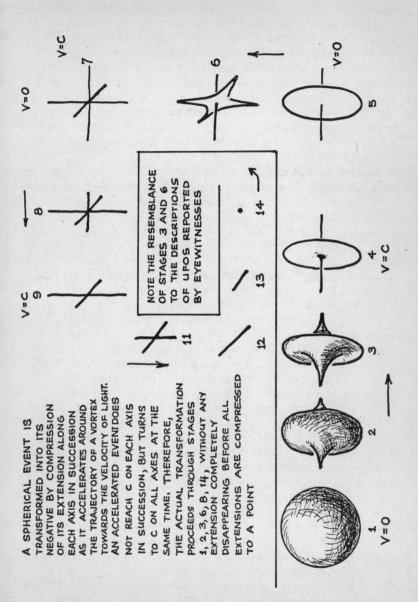

A SPHERICAL EVENT IS TRANSFORMED INTO ITS NEGATIVE BY COMPRESSION OF ITS EXTENSION ALONG EACH AXIS IN SUCCESSION AS IT ACCELERATES AROUND THE TRAJECTORY OF A VORTEX TOWARDS THE VELOCITY OF LIGHT. AN ACCELERATED EVENT DOES NOT REACH c ON EACH AXIS IN SUCCESSION, BUT TURNS TO c ON ALL AXES AT THE SAME TIME, THEREFORE, THE ACTUAL TRANSFORMATION PROCEEDS THROUGH STAGES 1,2,3,6,8,14, WITHOUT ANY EXTENSION COMPLETELY DISAPPEARING BEFORE ALL EXTENSIONS ARE COMPRESSED TO A POINT

NOTE THE RESEMBLANCE OF STAGES 3 AND 6 TO THE DESCRIPTIONS OF UFOS REPORTED BY EYEWITNESSES reconstructed the instant it passes through the Black Hole and the corkscrew trajectory unwinds, reversing the directions it took while accelerating. The information exploded into the quantum field re-forms the original standing wave, but now the acceleration is reversed. Time flows backward. We cannot see images of events flowing backward in time on the other side of Black Holes because all the radiation containing this information is flowing at the speed of light away from us. The past is reached by passing through a Black Hole. Whatever can be demonstrated by a tangible model can be translated into practical engineering.

As soon as explorers leave familiar landmarks by venturing over land or sea, they are in danger of becoming lost. As long as exploration was confined to the surface of a small planet, adventurers could wander all over the world with a fair chance of blundering their way back home. The advent of space exploration, however, demanded accurate courses plotted before the first Mercury capsule blasted off from Cape Canaveral; the navigator had to know exactly where to go and how to get back. Traveling to the future is not quite so simple as hopping into a time shuttle and going. Which way do you go?

If you blast off at the speed of light and return six months later to the location you left, all you will find is empty space. The Earth and the Solar System will be long gone in their galactic orbits, six months away. To reach a future Earth, you must plot a return course that will bring you to the location the earth will occupy at the other side of its orbit six months from now.

Relativity proves that the future lies in a straight line in any direction. The past extends from the line of travel at right angles in all directions. This is the basis of navigation through time. But if the future of this Earth lies only in the direction taken by this Earth along its trajectory in space, what future can possibly lie in all other directions? Each direction of travel leads to a different future. When Einstein demolished the concept of absolute time, he proved that each location in the universe travels a unique trajectory constituting a unique stream of time.

The reason time has been so refractory to scientific understanding is that we have conceived time to be a special kind of dimension. But time is not a dimension at all, any more than fire is the element the Greeks believed it to be. Time is measured in units of intervals, spans, durations, speed, and sequence; each of these measurements refers to a dimension of the waves constituting the universal field of vibrations. Einstein showed that a stream of time is defined by the unique sequence in which events are experienced at each location. Once Relativity proved that events follow each other in a unique sequence at each location, an infinite number of unique time streams must flow through the universe. Each time stream is identical to the course of travel in space.

The existence of alternative time streams can be deduced from a simple thought experiment. When you look at the Sun, the image of the Sun takes its place in your time stream in a certain spatial sequence relative to all the other events you experience. When you move, the position of the Sun relative to all other events in the universe changes in your perspective. Each relationship defines a different time stream. We are always moving through space. New time streams are created every time an observer changes directions. The creation of new time streams by steering courses in life is the basis of the concept of free will and an undetermined future. We are all spinning though space together on this planet, so the difference between everyone's personal time stream amounts to no more than a difference in the time of day when we look at the Sun.

Although sequence is created uniquely by each observer in the course of existence, events are eternal. This paradox can be explained by referring again to the Plate Flutter Model.

If the model is cut into the shape of a square and strummed along one edge, the powder will move from the edges to form a diagonal cross. The free vibration of the edges drives the powder to the diagonals, where there is no vibration at all. The edges represent a pole of pure radiant waves while the diagonals represent an opposite pole of pure standing waves. Between the two poles, the sonic velocity of the radiant waves is reduced along a line of continuous transformation toward standing waves. If a chord comprising a full range of musical notes is used to activate the model, the harmonic interactions of the different notes generate a sequence of audible undertones, and the patterns formed by the powder will move to correspond to the continuously changing harmonic beats. When the harmonic interactions exhaust all permutations of tone and sequence, the entire pattern in time and space will repeat itself again. This means that all the events the field of vibrations can possibly generate must exist somewhere in the field at all times as a wave pattern forming or disintegrating between the poles of pure radiation and pure standing waves. An event manifest as a standing wave represents a material body existing in the immediate present. All the wave patterns proceeding from the radiant pole toward the standing-wave pole constitute events of the future. All the wave patterns proceeding from the standing-wave pole toward the radiant pole constitute events of the past. In other words, everything that ever existed, everything that exists, everything that ever will exist, and everything that can possibly exist exists right now somewhere in the universal field of vibrations as a wave pattern in a state of continuous transformation.

Not only does every possible event exist at all times, somewhere in the field, as a wave pattern in a state of becoming real or disintegrating, but every event also exists everywhere throughout the entire field as a wave pattern in a state of realization or decomposition. To illustrate this deduction, waves reaching us from the Sun are manifest as a standing-wave structure when they come to a focus to form the Earth at this location. But radiation does not change from pure waves to pure matter the instant it reaches us. Over the entire distance from the Sun, wave patterns are building up the structure which will become the Earth when the radiation arrives here. The pattern of what the Earth will be in eight minutes' time already exists 93 million miles from us in space. By measuring the alignment of waves in the ambient field wherever we are, it is theoretically possible to calculate the location of any event in the universe and the time at which it will become manifest as a material body at that location. This is how a navigator plots courses through time.

Space is filled solid with vibrations. If all possible events exist as a wave pattern throughout the universe, it means that the space in your room must be filled with incompletely formed patterns of yourself doing whatever else you could be doing if you were not reading this. When you put this down and move, the standing-wave structure of your body disintegrates at the location where you are, here and now, and reforms in a succession of proximate locations along the path of your movement. But when you move, the wave pattern of you reading this continues to exist in an unmaterialized state. Outside of your home, patterns of you doing everything you can possibly do cover the entire Earth, not to mention other patterns of the Earth filling universal space. This is the basis of the science-fiction theory of alternative realities. Each sequence of events relating all possible events constitutes parallel time streams. Given enough time, all possible events must become materialized and all possible sequences must become realized. On another time stream, the Confederate States of America is defeated by a Soviet-Nazi Axis in World War II. I'd like to know where I missed the intersection with the time stream where I am irresistible to nubile blondes; that is probably the time stream where I become allergic to penicillin. This may not be the best of all possible worlds, but it isn't the worst, either.

Pure wave energy is not a mathematical abstraction. You can see it with your eyes. Each wave is seen as a light of the color defined by its frequency. Of course, we can see no more than a small spectrum with normal vision, but our eyes are sufficient to establish the principle. We cannot see a standing-wave structure because all material is a perfect black body. What we see is radiation emitted from either black bodies or white bodies. Between these two poles, radiation carries information of its source in the form of wave patterns. This information is not a mathematical abstraction either, because you can see it, too. Wave patterns between pure light and pure material exist as images.

Images which can be brought to a focus are called real images. We can see real images because our eyes can bring the waves to a focus. Images which cannot be brought to a focus are called virtual images. We cannot see virtual images. The space in your room is filled solid with wave patterns of yourself doing all the things you could possibly do if you were not reading this, but you cannot see these patterns because they are virtual images.

The reason virtual images cannot be seen can be demonstrated by looking at the Sun; don't burn out your retinas. You can see the real image of the Sun only along the radial line of sight. There is another image of the Sun just beside your head, but you cannot see it because the radiation does not impinge upon you. As long as you are where you are, the images of the Sun filling all the space around you cannot be seen. The other side of the Sun can never be seen from here because all the radiation is going away from us at the speed of light. The images of the Sun on all sides of you are virtual images on your time stream. Alternative realities exist in the quantum field as virtual state. All radiations in the virtual state of the quantum field possess a vector at right angles to our line of sight. This vector reduces the velocity of waves radiating toward us so that it no longer has the velocity of light at the point of impact. Therefore, no electromagnetic interaction can take place. Whatever does not interact with us does not exist as a real substance.

The solution to the Michelson-Morley paradox is *not* that light always travels at the same speed regardless of the relative velocities of the source and observer. 'Light' waves travel at all velocities conforming to the classical Newtonian laws of relativity—but only the waves traveling 186,000 miles per second can be detected as light. Waves of all other velocities are not detectable, so they disappear into the virtual state of the quantum field as corresponding waves 'materialize' from the cosmic plenum to maintain the absolute velocity of light.

A lens makes it possible to bend the flow of radiation so that virtual images are transformed into real images, making it possible for us to see parallel time streams. If you will take the lens of your camera—or a simple reading glass—to cast an image onto a small screen, you will make another discovery about the structure of the universe that will demolish your conceptions of reality and contribute further to the technological achievement of time travel.

Our cultural concepts lead us to perceive the image cast by a lens as two-dimensional. This is so self-evident to us that we cannot understand how anyone could possibly see anything other than a plane image. The image cast by your lens is really three-dimensional. You can prove it by moving the screen along the axis of projection to reveal an infinite number of planes of focus.

The normal lens transmits an image from an acceptance angle of 45 degrees square. Fish-eye lenses accept radiation within an angle of 180 degrees square. In theory, an ideal lens can transmit an infinite spectrum; a pinhole functions as this kind of lens. If you can imagine your lens as a fish eye with perfect transmission, you can see that within the image cone it projects, there is all the information required to reconstruct a replica of everything in the universe on the objective side. But light goes through a lens both ways. On the other side of the lens there is an image cone projected containing all the visual information pertaining to the other half of the universe. As most of this information is contained within a sphere you can cup in your hands, you can see that you can literally hold the entire universe in your hand. That is an awesome lot of information to pack into a handful of space.

You may remark that the image cone does not contain all the information needed to reconstruct everything in the universe because most of the universe is hidden behind obstructions to the line of sight. But no material structure is absolutely opaque. Radiation streams through everything. Material opaque to our eyes merely transforms the incoming radiation to invisible wavelengths. With proper technology, the radiation emitted by an opaque object can be transformed back to visible light so that we can see through the opaque material. Image amplifiers used on night scopes are examples to prove that your lens does, in fact, project all the information of the universe into its image cone.

Cultural concepts lead us to perceive that the information projected by a lens is received from a more or less distant source. If you know enough about photography to use filters, you know that all the information contained in the image cone existed a microsecond earlier at the surface of the objective element. In other words, all the information needed to reconstruct the entire universe exists at the surface of your lens, whether or not it is brought to a focus by projection into the image cone.

Now, move your lens around. You will find that no matter where the lens is located, it always projects a pair of image cones containing all the information in the universe. You have just proven that in the real universe, every part contains all the information of the whole universe. A structure that contains all its information in all its parts is a hologram. The universe manifests the structure of an infinite hologram.

In a universe possessing the structure of an infinite hologram, every possible event not only exists in material form somewhere at some time, but every possible event must exist everywhere all the time in a material form. In other words, the universe contains an infinite number of three-dimensional material structures in the same location at the same time. This can be possible only if the universal hologram extends into a fourth dimension of space. Space is not filled solid with dense material because all standing waves except the ones we can see and touch exist in the virtual state of the quantum field. The quantum field, therefore, is the fourth dimension. All events that are going to happen and all events that have happened are invisible and intangible to us because they exist in the virtual state; this is why Einstein described time as a fourth dimension. Actually, the time Einstein referred to is a fifth dimension extending through the quantum field; the fourth dimension to be discovered was energy level defined by frequency. When Max Planck discovered the quantum effect, he discovered a fourth dimension without knowing it.

The photographic lens cannot show you that virtual images of past and future events exist right here and now; much less can you see them exist in a material state. But another simple model will demonstrate the principle. Cut a strip of paper about an inch wide and two feet long. Put a half twist into its length and paste the two ends together. This construction is called the Möbius Strip. The Strip is a model of plane space extending for an infinite distance while rotating around 360 degrees on one axis and 180 degrees around the other. The Strip appears to have two sides and two edges, but if you trace a line with a pencil along one side, you will return to your starting point after traveling 720 degrees of length without crossing over an end or an edge. Likewise, if you trace along one edge, you will find that you have covered 'both' edges without raising the pencil from the paper. The Möbius Strip has only one side and one edge.

Now, if you trace a line around 360 degrees of length, you will find yourself back at your starting point but on the 'other side' of the paper, a paradox because there is no other side to a Möbius Strip. If the Strip were an ideal plane, the back side and the front side must not only be the same side, but they must also be in the very same space. If you cut a small asymmetrical figure— such as the letter L—from a piece of paper, and move it around the Möbius Strip for 360 degrees to reach the 'opposite' side, you will see that its orientation has been reversed up for down. In other words, traveling to the farthest end of a Möbius Strip is identical to remaining where you are and rotating the structure of your body 180 degrees on a horizontal axis.

Regardless of its length, any location on a Möbius Strip can be defined by a distance expressed as degrees around its circumference. If the length of the Möbius Strip is reduced to zero as a limiting case, it means that traveling to any location on its length is identical to rotating the equivalent number of degrees around a horizontal axis without changing location at all.

A universal hologram must possess the structure of a Möbius Strip; the only difference is that universal space rotates around five axes instead of two. This means that the farthest end of the universe occupies the same three-dimensional location as your material body does right now.

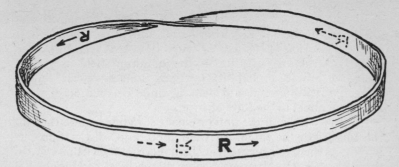

THE MÖBIUS STRIP

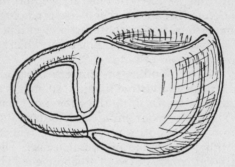

THE KLEIN BOTTLE

A standing-wave structure is the black-body pole of the universal field. The opposite pole is pure radiant energy. If our material body is what exists here, then pure light must be found at the farthest end of the universe where time ends.

But both poles of the universe exist in the same place, separated only by axial rotation of wave flow. Therefore, each person's body must be filled with light. This light can be seen in certain states of consciousness—it is the human aura.

The rotation of wave flow from the radiant pole is what transforms pure radiation into wave patterns and wave patterns into solid material and vice versa. In between the two poles of light and darkness, the wave patterns pass through stages of radiation as real images. These images are what we see radiating as photons from all objects. This radiation is the only way that black bodies can be seen.

It is the rotation of the line of sight that makes it impossible to see images of the Sun radiating on all sides of us. Therefore, it must be rotation of the axis of wave flow that transforms the material structures existing in the quantum field as virtual images of infinite probability into real images we can see and then into real material and back again. The rotation of wave axes is identical to the flow of time.

If the two poles of the universe coexist in our bodies, then all the images in the quantum field must also exist within our bodies. As these images are transformed from the virtual state to the real state, they become visible as the form of ideas in the mind.

The farthest end of the universe is the ultimate goal of all exploration. Now that we have reached it, we find that it has been here all the time. Because the universe is a hologram, we also find that our personal bodies contain a perfect replica of the entire universe. This is what William Blake saw when he discovered the universe in a grain of sand.

Outside of our personal bodies, the universe is perceived only as images—real images and virtual images. All images are replicated in the human body, forming the stream of ideas we call consciousness. Therefore, your mind already contains all the knowledge informing the entire universe. The real images in your mind constitute your state of consciousness. The virtual images in your mind constitute your unconscious. The whole universe is a state of consciousness, and the quantum field is identical to mind. The dichotomy of mind and matter, one of the basic conceptions of this civilization, is resolved. Although the discovery that the quantum field is identical to the mind blows the heads of Western scientists like a dose of acid, the fact that everything is a creation of mind is the basic concept of Oriental civilizations. This is why Orientals perceive space, time, and conservation differently from the way we do, and they developed a different technology. We call them backward mystics because we cannot comprehend that what they say and do are real.

The conclusion is inescapable. Time travel is a head trip. In fact, all transportation is a head trip. The Oriental mystics have always been telling us that *everything* is a head trip. It was this understanding that led them to train their minds to achieve the powers we call Psi, clairvoyance, telepathy, visions of the future, memories of past lives, levitation, soul travel, and the miraculous materializations commemorated in the miracles performed by Jesus. All that we disbelieve as mysticism is another kind of science, and all that we call science is our kind of magic. Whatever can be done by a mystical transformation of the human mind can be duplicated and made more reliable with technological assistance. This is why drugs are useful to stimulate the psychedelic experience, and why television has replaced clairvoyance, even in Japan.

So now we can understand that the achievement of scheduled time travel devolves on giving technological stimulation to the mind of the traveler. Chemicals have the troubles associated with blasting the mind in all directions at once. Television is limited to network scheduling. What we need is a portable TV transceiver in every garage capable of focusing on any location in the quantum field.

Note well that the camera lens reverses the orientation of the projected image—up for down, right for left, and proximity for infinite, like traveling around a Möbius Strip. The virtual state of alternative realities in the quantum field of time is the *negative* of material reality.

The lens shows us that it is possible to bring virtual images into focus as real images. The time transport, therefore, will be modeled after the operation of a fishbowl lens, capable of rotating waves flowing from all directions so that they come to a focal point at the intersection of three axes to materialize as a real structure. The principle has already been proven for the practicability of this kind of lens, beginning with the experiments of Nikola Tesla. The best-confirmed proof for transforming information from the quantum field to this reality is provided by the inventions of Henry Moray. An array of electrical transformers amplifying the geometry of a crystal lattice will function as a variable-focus zoom lens to rotate the flow of waves in a resonant chamber contained within its coils. David John Graham of Toronto has already put an electronic head transport on the market in competition with street drugs; his invention stimulates a psychedelic state by rotating its passenger in an electromagnetic field. It can be speculated that the Mark IV flying saucers use the electromagnetic field generated by the annular particle accelerators as a tunable transformer to bring about its levitation, flight, dematerialization into a holographic image, and materialization at other times and place.

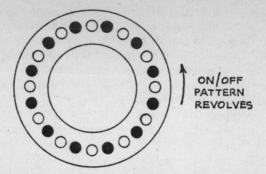

ON/OFF
PATTERN
REVOLVES

IF LAMPS IN A SERIES AROUND A CIRCLE ARE WIRED SO THAT EACH ONE SWITCHES ON AS SOON AS IT "SEES" THE ONE NEXT TO IT GO ON, AND SWITCHES OFF AS SOON AS IT "SEES" THE ONE NEXT TO IT GO OFF, THE PATTERN OF LIGHT WILL REVOLVE AROUND THE CIRCLE AT A SPEED CLOSE TO "C".

PATTERN FLOW ⟶

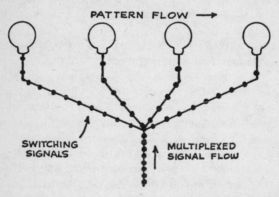

SWITCHING
SIGNALS

MULTIPLEXED
SIGNAL FLOW

SWITCHING SIGNALS CONDUCTED TO A SERIES OF LAMPS SO THAT EACH ONE GOES ON AFTER THE ONE NEXT TO IT, BUT BEFORE IT CAN "SEE" THE LIGHT OF ITS NEIGHBOR WILL CREATE A PATTERN CHASING AROUND THE CIRCLE FASTER THAN LIGHT, SENDING INFORMATION BEYOND THE "C" BARRIER OF RELATIVITY. BY REPLACING THE LAMPS WITH ARMATURES, AN ELECTRO-MAGNETIC FIELD CAN BE MADE TO SPIN FASTER THAN LIGHT WITH NO MOVING MECHANICAL PARTS. REPLACEMENT OF THE MAIN ROTOR AND GIMBAL GEARS WITH SOLID STATE ELECTRONICS CONVERTS THE MARK III FLYING SAUCER INTO A MARK IV TIME TRANSPORT.

Optical engineering has already proven the principles involved in the engineering of a variable-focus transformer with an acceptance angle of 360 degrees solid. When you look at the Sun, you are seeing rays diverging from a point source, and the image you see contains the past of the point source. Therefore, the past is found on diverging radiation. A converging lens is used to bring diverging radiation to a focus as a real image. When real images projected from all directions are brought to a common focal point in the resonant chamber contained by the transforming coils, a real material structure is created. The process is the reversal of the operations that make an atomic bomb explode. Conversely, radiation coming to you from all directions contains the information of your personal future on your unique time stream. Therefore, the future is found in the ambient field of the universal hologram by a lens that selectively filters the converging rays of a specific angle and bends them apart by a diverging optical formula. Diverging radiation contains virtual images, so another converging lens must bend the incoming radiation again to bring the virtual images into focus as a real image. When the real images are brought to focus from all directions, the future reality becomes materialized in the resonant chamber. Outside the resonant chamber, the transformers alter the angles of radiation from the capsule perceived by observers so that the time transport is seen to condense into the future or explode into the past, disappearing from detection as well as view.

Time is a function of field acceleration. Therefore, any engineering that can accelerate the acceleration of any field will serve as a time transport. If, for example, you can build a vehicle that will be accelerated by a gravitational field, you will eventually reach the velocity of light and disappear through a Black Hole. We already know that time travel can be achieved in this manner, and we already have a vehicle that is being accelerated in a gravitational field. We are traveling through time on the Spaceship Earth. What we want is to accelerate the gravitational acceleration so that we can get to the future sooner or later than everyone else in this planetary time stream.

As Einstein proved, a traveler can reduce the velocity of his personal time stream and reach the future by increasing the gravitational acceleration generated by a vehicle containing him. Conversely, a traveler can reach the past by reducing the attractive forces holding his personal material together;

reducing cohesive forces is identical to achieving hyperlight velocity so that matter waves explode into the quantum field. Time travel, therefore, is implicit in the antigravity engineering of flying saucers.

The most clearly defined experiment proving the feasibility of field engineering by the tuned focus of ambient radiation was performed by directing the beams of four microwave generators to a resonant chamber in the center. A hundred-gram weight suspended over the focal point loses seventy-five milligrams. A reduction in the acceleration of gravity can be expressed as a negative acceleration added to the universal constant. A negative acceleration of gravity carries a vehicle into the past. This experiment shifted the hundred-gram weight into the past about as far as orbiting astronauts are shifted into the future, but the achievement is as significant as splitting an atom of uranium for the first time. It is only a matter of time before enough engineers will have undergone a change of basic concepts of reality so that the manpower required to field a flying saucer, complete with its operational infrastructure, is assembled.

If time travel is feasible, then it must be inevitable on some time stream or other. If time travel is achieved on our time stream, then it must be achieved in our future because we have no record of the technology being developed in our past. If a future people have achieved time travel, we should expect an expedition of future historians to have visited us by now on their field trips into primitive civilizations. As we know no future anthropologists resident in our camp, it is reasonable to conclude that time travel was never achieved in our future; the theory will never be developed beyond speculation.

But time travel is not so simple as Margaret Mead taking a trip to New Guinea. Our belief about reality determines whether any records of visitors from the future will be entered into our annals. A lot of people testify that they have seen flying saucers, and a lot of photographs have been published. Nevertheless, all of the evidence and testimony have been denied credence by the authorities. I have little doubt that most of the reports are either mistaken identity or hoaxes, but even if they were true, the authorities would still deny the facts. Unless the UFOnauts decide to announce their presence to everyone by means of their own superior communications technology, we could be a colony of a galactic empire and never know it; the aborigines of the East Indies were colonies of the European nations for centuries without knowing that white men existed.

The beliefs a person holds determine what he will perceive of reality. According to Magellan's log book, cited by Lawrence Blair in *Rhythms of Vision*, the barefoot natives of Patagonia could not see the European ships when they arrived at South America for the first time. To the aborigines, the shore party appeared out of thin air on the beach. Eventually the shamans discerned a faint image of the tall ships riding anchor offshore. They pointed the images out to their tribespeople, and after everyone concentrated on the concept of giant sailing ships for a long time, the galleons materialized. And then the aborigines were annihilated by cultural shock. The UFO phenomenon is presenting our civilization with the same kind of experience.

Cruising the seas of time in a Mark V Flying Saucer is a profoundly different reality from sailing a Spanish galleon. When the time transport lifts off its launching pad on a trip to the past, it rises straight up on an antigravity beam and explodes into virtual radiation. The virtual radiation comes to a focus to form real images at the destination. The formation of a real image is attended with electromagnetic effects in the focal space; this is why disturbances are manifest where UFOs appear. Although a saucer from the future appears as an image to us, the vehicle is perfectly solid to its crew; it is we that appear to be a ghostly image. As a holographic image, the Saucer can maneuver without being subject to inertial forces generated by their incredible accelerations.

As an image, the Mark V is actually a mental structure. Whether or not one of us natives will be able to perceive a flying saucer depends on whether that person's consciousness is tuned to the proper time stream. All of us do not see the same material structures in this reality, as you realize in court; much less do we all perceive the same real images. What is a tangible body to some people is a real image to others. By and large, there is general agreement within any given civilization as to which patterns are solid material and which are merely images. The Patagonians, for example, perceived the European vessels to be images when they were first able to see them, while other people will perceive the images on a movie screen as material bodies.

At the borderline between real images and virtual images there is much less agreement as to which is real. When you experience a hallucination, the images you see are perfectly real to you, but they are merely virtual images to everyone else. If someone else could see your virtual image too, he would not regard it as a hallucination. Other cultures have a whole world of images which they recognize as real—the colony at Findhorn, for example—but those images are virtual to us, so we say that the whole community is subject to mass hallucination. The Mark V Flying Saucer travels along flight corridors where virtual images are transformed into real images. These corridors are the creation of the universal field as radiant energies are transformed into terra firma. Along these corridors, some people will see flying saucers, and other people won't. Some cameras will register a real image on film and others won't. An image will appear on some wavelengths while being absent on others. As long as the basic concept of our civilization maintains that whatever is real must be equally real for all observers, and all the more equal to all laboratory instruments, the flying saucer must be ignored as unreal.

Modern Americans are not altogether unlike the barefoot Patagonians who were psychologically blind to the sight of a superior technology. There are historical records of time travelers among us, but they are not perceived even by the people who read them. The twenty-fifty chapter of Exodus was written over three thousand years ago. Regardless of the distortions accumulated over the millennia of successive translations, there is not the slightest doubt that the King James Version has been reproduced to the letter for over three centuries; antique copies are still available to anyone determined to establish the date of printing. Now, if you follow the directions the Lord gave to Moses for constructing the Ark of the Covenant, you will build yourself a spark-gap radio transceiver powered by a Leyden-type condenser, almost identical to the models Marconi began his experiments with. Shittimwood is no longer stocked, and the price of gold is a barrel of oil, but pine and aluminum foil can be substituted. The Ark of the Covenant has been recognizable as an electronic device for the past hundred years; finding a modern radio in ancient Judea is as significant as finding a flying saucer in modern America. As long as the first is ignored and denied, even when you have the palpable evidence in your hands, you can be reasonably sure that the other will not be recognized on sight.

Critics disregard the Ark of the Covenant as a coincidence. After all, many religious artifacts have electrical properties. But the Bible clearly states that God told Moses to use the Ark for communication with Him. The Ark of the Covenant was intended to be used as a radio. Any God worthy of His divinity communicates to His Chosen People by means of mystical visions vouchsafed to Holy Prophets, so how come the Lord who walked the Earth as a man had to equip Moses with CB?

Radio cannot possibly be developed outside of a society so widely spread that electronic communication is necessary to keep it organized. The description of the Ark of the Covenent in the Holy Bible is proof that a vast and technologically sophisticated society existed on this Earth over three thousand years ago, within historical time, without leaving any recognition of their existence.

. In the heart of the Great Pyramid there is a stone coffer said to possess the same volume as the Ark of the Covenant. Considering the precision demanded by the Lord in the dimensions of the Ark, it is difficult to accept the coincidence between the Ark and the coffer as chance. The Pyramid is a solid-state module in a worldwide radio-communications network capable of broadcasting and receiving intelligence throughout interplanetary space. The Ark of the Covenant has the features of a mobile remote unit in the broadcasting system.

As well as being a radio, the Great Pyramid is the most precise and durable clock known to exist. Navigation through the seas of time is impossible without a chronometer of absolute accuracy. The Great Pyramid could not be better designed to function as an unmanned radio beacon marking traffic lanes through the holographic oceans of the universe. If this is so, we may expect to find other pyramids on other planets, like the black monolith in *2001: A Space Odyssey*.

The most fascinating problem of time travel is the Time-Loop Paradox. What happens when a traveler from the future meets his ancestors and alters history? It is generally concluded that any alteration of the past will cause the future to disappear, leaving the time trekkers stranded. The Time-Loop Paradox is inevitable if time travel is ever achieved, so this is the main argument against the feasibility of the technology. The Time-Loop Paradox, however, is a figment of the concept of absolute time. As soon as time trekkers touch down in the past, they create an alternative time stream, running more or less parallel to the manifold time streams we call our own world time, but existing in the virtual state of the quantum field. When the trekkers want to return home, they beam themselves back to the

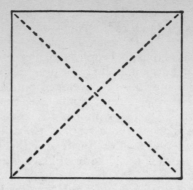

THE SQUARE AND ITS DIAGONALS ARE NEGATIVES OF
EACH OTHER.

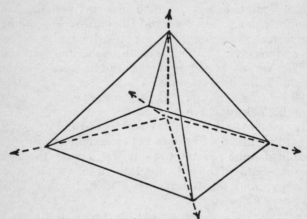

THE PYRAMID TRANSFORMS RADIANT ENERGY CONFORMING
TO ITS STRUCTURE INTO ITS NEGATIVE FORM AND BEAMS
THE INFORMATION ALONG THE NEGATIVE AXES.

future and then across the angular distance separating the two
divergent time streams. The proof that absolute time does not
exist provides a Time-Loop Exclusion principle that makes the
Time-Loop Paradox impossible.

The beliefs a society holds about reality determine the technology it will develop. The technology determines the direction
and the range the civilization will develop. As long as beliefs are
constant, technology proceeds slowly and steadily to the limit
of belief. When the limit of belief is reached, the development of

technology must stop. At this point, the civilization becomes old and decays. But all people do not accept the same beliefs. Those who do not agree with the cultural limitations are regarded as mystics, if not disregarded as insane. The leaders who take technology to the limit of the culture are the people in the most favored position to carry technology beyond the limits of belief. Advanced physicists have reached the limits of our civilization right now, and this is why the rest of us in the rear regard them as reverting to mysticism.

The insights and discoveries of the scientific vanguard cannot be believed by the people in the mainstream of civilization, nor can the facts make sense because the new information contradicts the basic concepts holding the civilization together. Taboo prohibits thinking about the basic concepts.

If it should ever happen that a party acquires the means to advance technology beyond the limits set by the culture, new concepts drive engineering to a quantum leap beyond the ken of ordinary people. When the force of this engineering becomes too great for the mainstream to deny, like the atomic bomb was introduced to the citizens of Japan, the civilization is subjected to cultural shock. The shock forces social leaders to examine the basic concepts of reality; the taboo is broken. When the taboo is breached, the civilization must collapse. Although it was the Japanese who were the target of the atomic bomb, Americans are also being subjected to cultural shock from the advance of technology, so the civilization of the victors is doomed to dissolution along with the vanquished. The only way to survive the collapse of civilization is by reconstructing the basic concepts of reality capable of comprehending the facts of life as rapidly as the old beliefs are swept away. Those who are capable of reconstructing their beliefs suffer a psychological death and rebirth into a new reality; they transfigure their civilization as a consequence of their personal renaissance and they become the founding fathers of a new civilization on the ruins of the old.

The development of any system of beliefs by expression through a technology fertilizes the seeds of doubt, which must eventually transform belief. The technology of Rome led ineluctably to the monotheism of Christendom. Unfortunately for the Romans, most of them were unable to change their beliefs before collapse set in, so little remains of Rome except ruins and an alphabet. The monotheistic belief of Christianity made it possible for an agrarian society to become transformed into an industrialized society. Industry brought belief of God into question by bringing the power of machinery into existence. The leaders of the industrial society changed their beliefs just as rapidly as technology advanced beyond the limits of the writers of the Holy Bible, so when Sir Isaac Newton appeared to lead civilization into a new age by proclaiming a Third Testament of classical physical laws, the transition was made without the usual traumata of death and rebirth. Albert Einstein is the latest prophet; he gave us a Fourth Testament of universal laws to guide us into the coming holographic age. The Japanese civilization has built-in feedback mechanisms enabling them to examine their basic concepts of reality at the drop of an atomic bomb, and reconstruct new concepts without dying from shock.

The industrial development of time travel, therefore, will be subsumed into the systems familiar to us, with no more changes than the technology enforces for its operation. Time travel begins with the testimony of prophetic visionaries who are disregarded as insane. But as soon as a few solitary freaks produce convincing proof that time travel is really possible, think tanks begin to organize around the world to do research, completely unknown to the general public. Then the military takes over and nothing more is heard for a generation. After all the mistakes and expenses are underwritten by the taxpayers, the machinery is handed over to Pan Am and Aeroflot to operate at a profit.

Time travel will be sold to the public when trips can be packaged like Cook's Tours. It would be impolitic to shatter faith in established reality by opening alternative time streams; commercial transport will be limited to the current time stream, and all other channels will be reserved for military patrol and exploration. You will buy your ticket for a vacation on a planet in the Sirian System much as you go to a travel agency for a trip to Hawaii. Departures are scheduled by harmonic windows through hyperspace, as sailing ships wait for wind and tide. All seats are coach class because transit is instantaneous. As soon as the mooring lines are cast off, beams of radiation come to a focus on the starship, and it dematerializes while accelerating straight up to the velocity of light. No one plays holographic chess to pass the time en route, because rematerialization at Starport follows immediately after takeoff. After docking, passengers are discharged. The crew takes leave to exotic fleshpots, cargo is off-loaded, and the officers prepare ship's articles for the return voyage. The return of the starship to Earthport could be scheduled for any time before or after the original departure, but there would be continual problems if tourists returned home before they left, and if they remained away more than a month, their employers would complain. Two weeks is just about the time everyone needs to complete the paperwork between shipments. The only limit on the time that travelers spend on the Sirian planet is the amount of money they take with them. The one problem defying solution is negotiating scale with the Space Jockeys' Interplanetary Union for time worked in hyperspace.

The holographic civilization is based on a concept of infinities. Infinity is supposed to be beyond any human conception and apprehendable only by God. But it isn't. It is our cultural beliefs alone that make it impossible for us to conceive of infinity. As holograms become increasingly understood, the concept of infinity will become as self-evident as any other basic belief, and we shall take another faculty for granted, a faculty hitherto perceived as an attribute of divinity.

Holographic concepts do not merely advance technology. The achievement of time travel literally demolishes the cosmos and creates a new Earth under new heavens. When it is possible to travel to the future, time ceases to exist as we know it. The future becomes an immediate present, just as Los Angeles is the suburb of New York to the jet set. The power to travel through time is the power to create sequence at will. Sequence is the principle of order by which all reality is established. Therefore, the technology of time travel confers upon the technocrats the power to create entire worlds to order. The engineers accede to the power of the immortal gods, transcending time and space.

When we meet the gods, we shall find that they are us.

THE END

INDEX

Prepare yourself for the future with–

THE BOOK OF PREDICTIONS
by David Wallechinsky, Amy Wallace and Irving Wallace

The creators of the bestselling BOOK OF LISTS and BOOK OF LISTS 2, have now consulted the leading minds on earth today, for their opinions on what will happen to us in the years to come . . . People like Arthur C Clarke, Milton Friedman, Alvin Toffler, Shere Hite, L Sprague de Camp and many others.

Read for yourself their startling new predictions, plus–

The 18 greatest predictors of all time

The 6 greatest predictions of all time

The worst predictions of all time.

Discover what psychics and seers predict for celebrities such as Brigitte Bardot, Bob Dylan, Richard Nixon and Jacqueline Onassis.

THE BOOK OF PREDICTIONS – Find out what's in store for us all!

0 552 11885 0   £1.95

THE BOOK OF LISTS
by David Wallechinsky, Irving and Amy Wallace

A truly unique compendium of offbeat learning and fun!

The 10 worst films of all time
7 famous men who died virgins
10 sensational thefts
15 famous events that happened in a bathtub
23 of the busiest lovers in history
9 breeds of dog that bite the most
10 doctors who tried to get away with murder
The 14 worst human fears

Plus much, much more!

Hundreds of lists on every subject imaginable involving people, places, happenings, and things, with biographies, nutshell stories and lively commentary throughout. Includes lists specially prepared for this book by Johnny Cash, Bing Crosby, Charles M. Schulz, Pele, Arnold Palmer and many others.

0 552 10747 6   £2.50

THE DIRECTORY OF POSSIBILITIES
edited by Colin Wilson and John Grant

Mythology and the Ancient World

The Occult and Miraculous

Strange Creatures and Unusual Events

Time in Disarray

Inner Space: Mind and Body

Outer Space: The Universe

The World of Tomorrow

In addition to these seven main areas of enquiry. THE DIRECTORY OF POSSIBILITIES includes an extensive alphabetical directory of minor topics, important personages and sites and intriguing improbabilities.

Much of the subject matter is by its very nature sensational and controversial. Some of the ideas put forward have never before appeared in print.

0 552 11994 6   £2.50

THE PROPHECIES OF NOSTRADAMUS
by Erika Cheetham

Four hundred years ago Michel de Nostredame sat alone in a dark, secret room studying the forbidden books on the practices of witchcraft and the occult. By his side stood a brass tripod and placed on that was a simple bowl of water. But the water shimmered and grew cloudy and from within its depths came visions of the past and the future . . . visions which told of the Great Fire of London, the Second World War, air travel, and even the assassinations of John and Robert Kennedy.

0 552 11567 3 £1.75

A SELECTED LIST OF FINE BOOKS
FROM CORGI

While every effort is made to keep prices low, it is sometimes necessary to increase prices at short notice. Corgi Books reserve the right to show new retail prices on covers which may differ from those previously advertised in the text or elsewhere.

The prices shown below were correct at the time of going to press.

☐	11992 X	**HOW TO SAVE THE WORLD**	*Robert Allen*	£1.50
☐	11567 3	**THE PROPHECIES OF NOSTRADAMUS**	*Erika Cheetham*	£1.75
☐	08800 5	**CHARIOTS OF THE GODS?** (Illus.)	*Erich Von Daniken*	£1.95
☐	09083 2	**RETURN TO THE STARS** (Illus.)	*Erich Von Daniken*	£1.95
☐	09689 X	**THE GOLD OF GOLDS** (Illus.)	*Erich Von Daniken*	£1.25
☐	10073 0	**IN SEARCH OF ANCIENT GODS** (Illus.)	*Erich Von Daniken*	£1.95
☐	10371 3	**MIRACLES OF THE GODS** (Illus.)	*Erich Von Daniken*	£1.50
☐	10870 7	**ACCORDING TO THE EVIDENCE** (Illus.)	*Erich Von Daniken*	£1.25
☐	11716 1	**SIGNS OF THE GODS?**	*Erich Von Daniken*	£1.50
☐	10747 6	**THE BOOK OF LISTS**	*David Wallechinsky, Irving Wallace & Amy Wallace*	£2.50
☐	11681 5	**THE BOOK OF LISTS 2**	*David Wallechinsky, Irving Wallace & Amy Wallace*	£2.25
§	01137 5	**THE PEOPLE'S ALMANAC 2**	*David Wallechinsky, Irving Wallace & Amy Wallace*	£4.95
☐	11885 0	**THE BOOK OF PREDICTIONS**	*David Wallechinsky, Irving Wallace & Amy Wallace*	£1.95
§	01352 1	**THE PEOPLE'S ALMANAC 3**	*David Wallechinsky, Irving Wallace & Amy Wallace*	£3.95
☐	11994 6	**THE DIRECTORY OF POSSIBILITIES**	*Colin Wilson & John Grant*	£2.50

ORDER FORM

All these books are available at your book shop or newsagent, or can be ordered direct from the publisher. Just tick the titles you want and fill in the form below.

CORGI BOOKS, Cash Sales Department, P.O. Box 11, Falmouth, Cornwall.

Please send cheque or postal order, no currency.

Please allow cost of book(s) plus the following for postage and packing:

U.K. Customers—Allow 45p for the first book, 20p for the second book and 14p for each additional book ordered, to a maximum charge of £1.63.

B.F.P.O. and Eire—Allow 45p for the first book, 20p for the second book plus 14p per copy for the next 7 books, thereafter 8p per book.

Overseas Customers—Allow 75p for the first book and 21p per copy for each additional book.

NAME (Block Letters) ....

ADDRESS ....

....